Signposts for geography teaching

Papers from the Charney Manor Conference 1980

Edited by **Rex Walford,**
University of Cambridge

Longman

LONGMAN GROUP LIMITED
Longman House
Burnt Mill, Harlow, Essex

First published 1981
ISBN 0 582 35334 3 (cased)
ISBN 0 582 35335 1 (paper)

Set in 10/12 pt *Monophoto Times*
by Butler & Tanner Limited, Frome and London

Printed in Great Britain
by Butler & Tanner Limited, Frome and London

Acknowledgements

We are grateful to the following for permission to reproduce copyright material:

Chappell Music for an extract from lyrics of *Another Brick In the Wall* words and music by George Roger Waters © Pink Floyd Music Publishers Ltd. Administered by Chappell Music Ltd; The Journal of Philosophy Inc and the author, Lawrence Kohlberg, for an extract from pp 631–632 'The Claim to Moral Adequacy of a Highest Stage of Moral Judgement' *Journal of Philosophy* LXX: 18, October 1973, and Macmillan, London and Basingstoke, for the poem 'Geography Lesson' from *Jets from Orange* by Zulfikar Ghose.

Contents

List of participants
Foreword
Introduction *Rex Walford*

Part One: **Teaching Units**
Introduction 3
I The impact of the Industrial Revolution on an English city – a location exercise *Neville Grenyer* 5
II Hunters and collectors – a study in changing pupil perceptions of other societies *John Bale* 16
III Does development make you happier? – an exercise in quantifying the qualitative *Rex Walford* 22
IV Murderer at large! – a simple map exercise that introduces the computer *Peter Fox* 25
V Changing urban geography – using house prices as a data source *Andrew Kirby* 28
VI Fourth year urban fieldwork – hypothesis testing in Cheltenham *Derek Plumb* 33
VII Income distribution – an operational game *John Bale* 36
VIII Slopes, soils and vegetation – a fieldwork exercise in biogeography *Graham Corney* 40
IX Drawing boundaries around a conurbation – an exercise in 'new' regional geography *David Lambert* 45
X Teaching welfare issues in urban geography – a work unit on externalities *John Bale* 51
XI On teaching about values in the classroom – a lesson and a retrospect *Liz Ambrose* 63
XII More shops for Abingdon! – a decision-making exercise *Eleanor Rawling* 66

Part Two: **Appraisals of the 1970s**
Introduction 81
1 Three schoolteachers reflect on geography 1970–1980 *Sheila Jones* *John Rolfe* *Pat Cleverley* 83
87
91
2 Quantification and school geography – a clarification *Roger Robinson* 94

3 Examinations in the 1970s *Chris Joseph* 97
4 The overlap between school and university geography *Vincent Tidswell* 106
5 Seven reasons to be cheerful? ... or school geography in youth, maturity and old age *Andrew Kirby and David Lambert* 113
6 Geography and the school curriculum debate *Richard Daugherty* 119

Part Three: **Signposts for the future**
Introduction 131
7 Towards a human geography *Derek Gregory* 133
8 Geography and values education *John Huckle* 147
9 Progression in the geography curriculum *Trevor Bennetts* 164
10 Educational computing and geography *David Walker* 185
11 Towards a new generation of teacher-technologists *Michael G. Day* 191
12 New opportunities in environmental education *Eleanor Rawling* 203
13 Evaluation in the 1980s *Gerry Hones* 213
14 Language, ideologies and geography teaching *Rex Walford* 215

Charney Manor Conference, February 1980

Participants and their posts at the time of the conference

LIZ AMBROSE	Priory School, Lewes, Sussex
JOHN BALE	Department of Education, University of Keele
TREVOR BENNETTS	HMI (Staff Inspector in Geography)
MARGARET CAISTOR	HMI
CHRIS COLTHURST	Lancing College, Sussex
GRAHAM CORNEY	Department of Educational Studies, University of Oxford
RICHARD DAUGHERTY	Department of Education, University of Swansea
NEVILLE GRENYER	Winchester College
GERRY HONES	Department of Education, University of Bath
JOHN HUCKLE	Department of Geography, Bedford College of Higher Education
SHEILA JONES	Colston's Girls' School, Bristol
CHRIS JOSEPH	Marlborough College, Wilts
ANDREW KIRBY	Department of Geography, University of Reading
DAVID LAMBERT	Ward Freman School, Buntingford, Herts
DEREK PLUMB	Cleeve School, Cheltenham, Glos
ELEANOR RAWLING	Schools Council, 16–19 Geography Project
ROGER ROBINSON	Department of Curriculum, University of Birmingham
ROSEMARY ROBSON	Eastbourne Sixth Form College, Eastbourne
JOHN ROLFE	Haberdashers' Aske's School for Boys, Elstree, Herts
GRAHAM STEVENS	Haverstock School, Hampstead, London
LESLEY TUMMAN	Geography Adviser, Derbyshire
VINCENT TIDSWELL	Department of Educational Studies, University of Hull
REX WALFORD	Department of Education, University of Cambridge

Foreword

In 1970 a group interested in geographical education met for a weekend of discussion and for an exchange of lesson ideas at Charney Manor in Oxfordshire. The conference had no official status and was an informal gathering of innovating classroom teachers and lecturers who were involved in the general diffusion and development of some fresh approaches to the teaching of geography.

A set of themes which had come (somewhat misleadingly) to be called 'the new geography' was particularly stimulating at that time, and these themes were generally based around the belief that spatial concepts, enquiry and problem-solving activities, games and simulations, and structured work-programmes could be combined into a 'geography of ideas' rather than one theme which was heavily oriented towards the description of the world and the provision of factual information.

A book emerged, without premeditation, from that conference and *New Directions in Geography Teaching* (Longman 1973) served to diffuse many practical classroom strategies to a wider audience, as well as reflect some thinking about new developments in geographical education.

Ten years later most of the original conference participants again gathered at Charney Manor to discuss and compare their experiences of the past decade and to again try to look ahead. They were joined by another dozen teachers and educators who were closely involved with current teaching and curriculum initiatives.

Compared with 1970, the topics of discussion had widened considerably. It was also clear that there was a greater multiplicity of viewpoints. The presentations and discussions of that weekend form the basis for the contributions to this present volume.

Both an appraisal of the 1970s (Part Two) and a consideration of the future (Part Three) took place.

Central to the theme of both conferences was a desire to see what the impact of new ideas means in practical classroom terms; thus, in *Signposts*, as in *New Directions*, the first half of the book is devoted to examples of teaching units (Part One) which represent some of the interesting and innovative work being done in geography classrooms at the time of writing. The units lay no claim to being *typical* of current classroom practice.

In the same vein, the other contributions do not set out to make a complete survey of geographical education. If they had attempted to do so, there would be much more in the book about education for graphicacy, work in multi-cultural classes, the widespread impact of Geography for the Young School Leaver (GYSL) materials, the development of rational course planning and so on.

It is hoped, however, that these transcripts of papers presented at the conference – with the addition of three others subsequently added – introduce some issues which will provoke geography teachers and educators to thought.

Writing for the future is a risky business, since in today's world there is no way of telling that horizons even ten years away are in sight. To predict, through a study of present trends, may be wishful thinking. Nevertheless, the intention within these covers is to provide some signposts which may enable geography teachers and educators to explore and evaluate some of the interesting avenues which open up as they search for a route into the daunting world of the 1980s and beyond.

R.W.

FRASER: Well, I did. I couldn't bear the noise and the chaos. I couldn't get free of it, the enormity of that disorder, so dependent on a chance sequence of action and reaction. So I started to climb, to get some height you know, enough height to drop from to be sure, and the higher I climbed the more I saw and the less I heard.

And look now. I've been up here for hours, looking down, and all it is is dots and bricks, giving out a gentle hum. Quite safe. Quite small after all. Quite ordered, seen from above. Laid out in squares, each square a function, each dot a functionary.

I really think it might work. Yes, from a vantage point like this, the idea of society is just about tenable.

From *Albert's Bridge* (1969) by TOM STOPPARD

Geography Lesson

When the jet sprang into the sky,
it was clear why the city
had developed the way it had,
seeing it scaled six inches to the mile.
There seemed an inevitability
about what on ground had looked haphazard,
unplanned and without style
when the jet sprang into the sky.

When the jet reached ten thousand feet
it was clear why the country
had cities where rivers ran
and why the valleys were populated.
The logic of geography –
that land and water attracted man –
was clearly delineated
when the jet reached ten thousand feet.

When the jet rose six miles high
it was clear that the earth was round
and that it had more sea than land.
But it was difficult to understand
that the men on earth found
causes to hate each other, to build
walls across cities and to kill.
From that height, it was not clear why.

ZULFIKAR GHOSE

Introduction

Tom Stoppard and Zulfikar Ghose - neither of them geographers in the academic sense - present more graphically than any technical explanations the change of mood between 1970 and 1980 in geographical education.

Stoppard's character Fraser goes up on to Albert's Bridge in order to commit suicide; down below at street level he is confused by the 'noise and the chaos' by 'the enormity of that disorder'. Once up on the bridge, however, he looks down from a great height and discerns that there is pattern and ordered process. 'Yes, from a vantage point like this the idea of society is just about tenable.'

So it was for geographers, shell-shocked by the data explosion of the 1950s and 1960s, who sought to make sense of the informational noise and chaos. A climb to the eyrie of spatial theorists moved interest from the particular to the general, from the discussion of differences to the recognition of similarities. Like Fraser, they found the idea of society 'just about tenable' if templates of spatial patterns were applied. The filtering out of some information in the process of applying these templates was rather thankfully done as some semblance of order began to emerge in the world view; it reinvigorated and liberated traditional teaching strategies in the process, and provided an exciting new perspective on over-familiar material.

In 1970, when a group of innovating teachers and lecturers gathered at Charney Manor, many were under the influence of this new perspective. Earlier in the 1960s, the main themes and insights of this approach to geography had been introduced to teachers at conferences held each year at Madingley Hall, Cambridge. A subsequent set of conferences organised jointly by Her Majesty's Inspectorate and the Geographical Association (the professional subject body for geography teachers) at Maria Grey and Doncaster Colleges of Education were to maintain a considerable driving force for change.

This force was given added impetus by being harnessed to new insights and understanding which were provided by the writings of educational theorists at the time and being applied to geography teaching. The writings of Hirst[1] and Peters[2] in England caused geographers to speculate about the exact nature of their own discipline: if not a 'form' of knowledge, was it any less respectable for being a 'field', coherent in content, rather than as a 'way of knowing'? Was it possible to isolate what was *distinctive* about the fruits of a geographical education, identifying that which no other part of the curriculum provided?

Even more influential was the work of the American psychologists

Bloom[3] and Bruner[4]. Bloom provided a 'taxonomy of educational objectives' which provided a base from which to work out ways in which students were asked to think; it helped to dethrone 'remembering' as an important geographical skill, and place high-order thinking skills such as analysis, synthesising and evaluating as more important objectives.

Bruner's writing on curriculum development was all the more influential because of its succinctness and lucidity. His idea of the 'spiral curriculum' based on key ideas in a discipline rather than its content became a regular part of curriculum thinking in geography, and his book, *The Process of Education*, had an influence out of all proportion to its ninety-two pages.

It seemed clear then that geography was ripe for some kind of radical change; the restricted objectives and formal style of descriptive and informational teaching offered little satisfaction to young teachers who had found a very different kind of geography emerging in their university studies. Those who had bitten on the bullets of Chorley and Haggett[5,6], or who had struggled with the philosophy of Harvey[7] were keen to see if this could be translated into realistic classroom activities.

The American High School Geography Project (1963–70) was the first attempt to translate some of this material to secondary school level, and its units of work were influential among English innovators, although its selling price was far too high ever for it to have great commercial success in the UK. (The possibility of buying individual copies rather than class sets in a revised version, published in 1980, has gone some way towards changing this situation.)

The growth of 'spatial analysis' in universities seemed to offer for school geography the chance of a sharper, more thoughtful approach, a geography structured around ideas rather than facts. Concepts such as 'best location', 'diffusion', 'best route' and so on, could be drawn from the spatial domain and allied to a battery of new techniques which seemed seductively powerful, correlation and regression, network analysis and map transformations. The puritan landscapes of the isotropic plains and billiard-ball universe were literally a world apart from the capes and bays of more traditional study.

Some of these techniques depended on a basic numeracy, but to characterise the whole movement as 'the quantitative revolution' is a vast over-simplification if not outright falsification of the truth. Quantitative techniques were introduced in concert with some of the new work activities, but they were the catalysts and not the impulse for such work. They presented a convenient alternative language used to distinguish an ambitious quest for precision and accuracy from some of the simplifications and vaguenesses of past eras in geographical education.

These changes took place against a background of great upheaval in the educational system of England and Wales. The process of comprehensive schools replacing the old tri-partite system was in full flow; the school-leaving age had been recently raised to 16; there was an expansion in school numbers and as a result in the numbers of teachers in training. Thus, there was ample opportunity for the development of new courses and ideas in these uncharted waters; there was also scope for speedy promotion to positions of influence and responsibility in a greatly expanding educational economy.

The excitement of this time of change is clearly remembered by those who were at the first Charney Manor Conference, held in 1970. There had already been skirmishes with the geographical establishment and a rival to the GA came into being for a short time. An issue of *Geography* (January 1969) guest-edited by R. J. Chorley gave a first major public platform to exponents of these new ideas in geographical education, and by 1970 it could be perceived that many important barometers of opinion were swinging round in response to pressure.

J. W. Morris, then Staff Inspector for Geography in Her Majesty's Inspectorate, was one of those influential in encouraging and developing change. The view of HMI was symbolically transmitted when the official HMSO publication for geography teachers of the 1960s, *Geography in Education*, which featured a map of the historic Hereford 'mappa mundi' on its cover, was replaced in 1973 by *New Thinking in School Geography*. The latter, now to be the repository of 'official thinking', was covered with a pattern of Christaller-style hexagons.

Meetings and conferences helped to promote the important face-to-face contact between innovators, enquirers and academic frontiersmen and an 'underground press' of worksheets, pamphlets and informal magazines developed for a time, as interested parties swapped ideas.

After a necessary delay whilst classroom experimentation forged and refined ideas, there began to appear a stream of classroom and teachers' texts which presented interpretations of the 'new' material. Surprisingly, it was a primary school textbook series, *New Ways in Geography*, by J. P. Cole and N. J. Beynon (Blackwell), which led the way, but this was produced as the result of particular circumstances, since the senior author was a university professor keenly interested in the new academic climate on behalf of his young children.

Everson and FitzGerald[8,9] produced early seminal books which revealed the shape of new human geography in the sixth form. Vincent Tidswell[10] went a long way towards persuading teachers to try out new techniques with which they were not familiar. (It is significant and revealing that the sister book planned on *physical* geography techniques did not appear until 1980; it has been the human side of the subject which has largely made the running in the 1970s.)

It may be invidious to select lower school coursebooks as representative of the decade, but a best seller has been the *Oxford Geography Project*

(OGP), series (OUP), co-authored by John Rolfe and members of staff from The Haberdashers' Aske's School, Elstree, based on their own school course in years 1–3. Haberdashers' intake is selective in ability range and so the books have provided stimulus, but not practical help for some. They have engendered a large number of other texts, most aimed at a slightly wider range of abilities than the OGP, but bearing the same general pattern, i.e. a set of books with much work activity as well as content, based on systematic topics and themes, rather than on an encyclopaedic Cook's tour of a continent.

Though the use of a textbook has been characterised as a strait-jacket on innovation, there is no doubt that the steady provision of course-books by publishers has helped the spread of new ideas beyond the enthusiastic pioneer core. They have provided concrete examples of classroom strategies within a coherence of a course framework; Heads of Departments who did not train in the time of 'new geography' at universities have been quietly thankful to puzzle it out from the book provided for their students' work....

A further step in the institutionalisation of change – and perhaps the key one – has come through the involvement of the external examination boards. For many schools 'The examination *is* the curriculum' and therefore movement in this context flutters the sleepiest of dovecotes when all else has failed.

The Oxford and Cambridge Board led the way with a new A-level syllabus as early as 1968, but their clientele (from the Independent sector for the most part) was not altogether a representative one. Larger boards were more circumspect about change. The Joint Matriculation Board, one of the two largest, performed a complicated A-level minuet in which an alternative syllabus was eased alongside a traditional one to assuage radical demands, whilst the traditional one itself was undergoing revision; the London board – the other of the 'big two' – decided, after long consideration, to make a change at A-level without transitions. This caused alarm in the backwoods for a time, and in-service courses based on the work of a new syllabus were severely over-subscribed for a year or two.

Alongside these changes came the impetus provided by the work of two partly Government-funded Schools Council projects in geography – the 14–18 Project based at Bristol, and the Geography for the Young School Leaver Project, based at Avery Hill. The fruits of the Project teams' work became generally available in the mid-1970s.

The Bristol Project, given the brief to work with average and above-average 14 to 18-year-old pupils, managed to complete only the 14–16 part of their task. But, working with a relatively small number of pilot schools, they succeeded in putting a new O-level on the market which gave considerable teacher autonomy in syllabus planning and which allowed 50 per cent of the final examination mark to be internally assessed through coursework and an individual study. This gave unparalleled scope to teachers willing to assume the responsibility for both curriculum development and evaluation, and opened up the possibility of sensitively-evolved work continuously relating to external validation.

The Avery Hill Project, working with much more pressing needs in mind, produced some comprehensive kits of material (becoming known as GYSL materials) based on three themes, *Man, Land and Leisure*, *Cities and People*, and *People, Place and Work*, which were effectively marketed and disseminated to over half the secondary schools in the country. Their diffusion success – unparalleled by any other Schools Council major project – was a testimony both to the excellence of the materials, and to the need of many teachers working with average and below-average groups. The GYSL materials were not totally in the tradition of spatial analysis, nor were they totally within the discipline of geography, but the impact of new movements in geography can be easily discerned within them. Their concern, as stated in teachers' handbooks and in the presentation of materials, is with concepts, skills and values in geography, and they relegated the teaching of factual information to an adjunctive role.

The Bristol Project spawned CSE-equivalents just as the Avery Hill syllabuses inevitably became the subject of an O-level examination. These two consequences added extra variety to the plethora of new syllabuses which the other GCE boards were developing and which were complemented by the numerous other syllabuses of the recently formed CSE boards – now, in practice, examining pupils down to about the 80th percentile of ability. By the time 1980 arrived, no examination board retained the syllabuses at O- and A-level which it had used in 1970. In most cases, the impact of 'new geography' was apparent.

This case history of change does not, however, seek to ignore the fact that there were some teachers (perhaps even a majority) who viewed changes in school geography with suspicion. Some grumblingly changed, others sought administrative havens with relief, a third group tried to avert their gaze altogether. Acerbic criticisms of the 1970s were not always unjustified.

Some of the new examinations, developed by protracted committee negotiations, looked even fuller than before, even though structured in fresh-looking systematic or conceptual language; some ideas suggested for the classroom stimulated teachers more than they appeared relevant to children; the wilder excesses of academic concern were sometimes pursued too fast and too blindly by those anxious to maintain a link between the research frontier of the subject and its general school image.

And even amongst protagonists of the new ideas, there was a realisation that a commitment to evolution and change meant an uncomfortable fluidity within teaching contexts and syllabus development. The group who appeared to be marching with such vigour towards a Promised Land in 1970 were not all totally convinced by Canaan when they arrived. Some, having taken a quick tour, have decided to strike out on expeditions in other directions, feeling that travelling hopefully is better than arriving.

Some of that feeling is elegantly encompassed in the Ghose poem which forms the second of the quotations at the beginning of this chapter. 'When the jet reached ten thousand feet ... the logic of geography was clearly delineated ... but from that height it was not clear *why*.'

If spatial analysis had done much to revivify the study of geography (and in some sectors of education make it a 'respectable intellectual discipline' for the first time) it had also incidentally intellectualised it in a way which sometimes removed its colour and its charm. It had put geography in danger of losing its roots in the real world because of a preoccupation with normative models and theories. Such a danger was probably inevitable, but by 1980 the pendulum had begun to swing again.

Many of those who gathered at Charney Manor in 1980 felt that geography had changed much for the better in the intervening decade, but that other horizons now opened up and that it was fatal to remain on a plateau of consolidation for too long a period.

In particular, there was reiterated a desire to return 'humaneness' to geography; to make sure that on the one hand the subject considered the 'deep structures' of society in developing explanations, and on the other that it was committed to the integrating of both mind and emotions in helping to orientate students realistically to the world in which they lived.

Consequently, there was talk of house prices and movements of capital in city areas, rather than a consideration of classical economic bid-rent theories in a spatial context; there was concern about teaching committed attitudes towards inequality and poverty rather than to merely identifying it and commenting on it.

Not all conference members felt this was a practical way forward for the teaching of the subject in schools; some expressed the view that much more had to be done in order to profit from the gains of the past decade, and that it was still preferable to adopt a positivist attitude to the environment within the context of formal educational structures.

This view was often coupled with the cautionary outlook that education faced a difficult decade in the 1980s because of economic recession, and a fall in school population. The reduction in student numbers will be accompanied by a fall in the number of teachers, and of those being trained (see Table 1 for graphic evidence of this). Consequently change agents will be fewer, and we might expect to find less change and innovation in the next decade than in the last one.

Nevertheless, even the most cautious were anxious to debate the new signposts which the presenters of some papers afforded. Parts One and Three of this volume offer some of the basis of that stimulating discussion.

Table 1 *The changing output of trained teachers who have geography as the major component of their studies. Results based on a survey of institutions with initial training units, England and Wales (85% response to questionnaire).*

	1970				1978					1981				
	PG	*B.Ed.*	*Cert.*	*Total*	*PG*	*B.Ed.*	*Cert.*	*Total*	*% 1970*	*PG*	*B.Ed.*	*Cert.*	*Total*	*% 1970*
Primary	23	124	967	1 114	33	298	268	599	53	33	215	19	267	24
Middle	36	114	722	876	60	295	204	559	64	67	215	12	294	34
Secondary	438	152	498	1 088	551	331	128	1 010	93	610	222	0	832	76
Total	497	394	2 187	3 078	644	924	600	2 168		710	652	31	1 393	
% 1970 Total					129	234	27	70		143	165	15	45	

PG – Post Graduate Certificate – 1981 figure based on places available
B.Ed. – Bachelor of Education
Cert. – Certificate of Education

(*Source:* Survey undertaken by Teacher Education Section of the Geographical Association, 1979. Copyright to GA)

And, one is left to speculate, if geography teachers find themselves increasingly required to defend the existence of their subject in a utilitarian-influenced curriculum (as Richard Daugherty suggests in Part Two), there is every need for discussion about new ideas to continue, rather than to stagnate. Those who can give no reasons for their activity and assume that their subject's value to education is self-evident are precisely those most at risk.

While practitioners of geography maintain a healthy dialogue between themselves and while innovators continue to fly kites across its intellectual landscape, it will remain vibrant enough to respond to the challenge of those who consider it a useless or unimportant dimension of education.

References

1 HIRST, P. H. (1968) 'The contribution of philosophy to the study of the curriculum' in J. F. Kerr (ed.) *Changing the Curriculum*. ULP

2 PETERS, R. S. (1966) 'Education as initiation', *Ethics and Education*. Allen and Unwin

3 BLOOM, B. S. *et al.* (1956) *Taxonomy of Educational Objectives: Handbook 1, The Cognitive Domain*. Longman

4 BRUNER, J. S. (1960) *The Process of Education*. Vintage

5 CHORLEY, R. J. and HAGGETT, P. (1963) *Frontiers in Geography Teaching: the Madingley Lectures for 1963*. Methuen.

6 CHORLEY, R. J. and HAGGETT, P. (1967) *Models in Geography: the second Madingley Lectures*. Methuen

7 HARVEY, D. (1969) *Explanation in Geography*. Arnold

8 EVERSON, J. A. and FITZGERALD, B. P. (1969) *Settlement Patterns*. Longman

9 EVERSON, J. A. and FITZGERALD, B. P. (1972) *Inside the City*. Longman

10 TIDSWELL, W. V. (1973) *Pattern and Process in Human Geography*. UTP

Part One

Teaching Units

Introduction

Twelve teaching units make up this section, varying from single lesson plans to extended pieces of work designed for several weeks.

The units do not pretend to be *representative* of what happens in British schools in 1980, but have been chosen to illustrate some of the more interesting and innovative work which is going on in classrooms at the present.

In a comparable volume to this, gathering material from teachers in 1970, there was a considerable coherence and communality of approach in a parallel set of published lessons.

By contrast, the 1980 set represents a much more diverse set of influences, a situation which probably reflects with accuracy the current state of the discipline of geography.

Graham Corney (Unit VIII) and Derek Plumb (Unit VI) both record fieldwork exercises born from the quasi-scientific hypothesis-testing approach which made such impact on field teaching at the end of the 1960s and which has encouraged much investigative and self-motivated work in school geography since.

Andrew Kirby (Unit V) presents a quantitative exercise concerned with house-price data which recognisably derives from the spirit of the 'quantitative revolution' although its content and its implications reveal a shift from the early days of urban modelling in schools.

John Bale's income distribution game (Unit VII) and Eleanor Rawling's locational decision-making exercise (Unit XII) have their roots in the development of experiential activities which have permeated much geographical methodology in the last decade, as does Neville Grenyer's intriguing location exercise (Unit I); but two of the three have been extensively influenced by 'welfare approaches' to geography, and Unit XII reveals itself to be a full-blown examination in another guise.

The exercise about hunters and collectors (Unit II) has a deceptive traditional look about it, but its ultimate intentions are to draw attention to the stereotyping which we bring to viewpoints about other peoples. John Bale's other contribution (Unit VII) owes something to the concerns about perception which have been present in academic geography in recent years, and Rex Walford's exercise (Unit III) is in a similar vein.

Liz Ambrose brings in to the open (Unit XI) the question of teaching about values, an issue which has hovered on the fringes of school geography for some time; and Peter Fox (Unit IV) demonstrates that the microcomputer is a friendly and helpful tool in simple map-work, rather than merely a plaything for the *aficionados*.

David Lambert (Unit IX) seems to bring the wheel full circle with his exercise on drawing regional boundaries, but it is a more realistic and local region which he examines and not the pleasure-domes of the 1950s.

I The impact of the Industrial Revolution on an English city – *a location exercise*

Neville Grenyer

The comparable 1970 volume to this book – *New Directions in Geography Teaching* – began a series of 'Teaching Units' with Brian FitzGerald's 'Iron and Steel Game', as an example of an innovative and exciting new classroom approach.

In the intervening decade, that exercise (and several imitators of it) has made its way into many secondary school classrooms, often as a key activity in stimulating students to consider location factors and the influences upon them. It seeks to simulate the historic locational decisions of industry in a free-enterprise society, and its emphasis (in both real-world and classroom terms) is on making a 'profitable' choice of location.

Such a view of location has been the prevailing concern until comparatively recently. But Neville Grenyer's role-playing simulation, described below, significantly demonstrates a shift of emphasis in geography between the start of the 1970s and the start of the 1980s.

Its apparent initial concern with optimal location in the classic profit-making sense becomes transformed with deeper 'welfare issues' as the exercise proceeds. Questions of social equity and justice intrude upon the simpler assumptions of the initial model.

The exercise has been used at all levels from 13-year-olds to 18-year-olds, with the differences in treatment lying mainly in the depth and complexity of class discussion arising from the activity.

Introduction

This classroom exercise relies for its effect upon deceit. The exercise starts by pupils being given the roles of early-nineteenth century industrialists competing with each other for the best locations in Manchester. However, when these sites have been chosen, the profitability of the various locations is ignored and instead the class concentrates on an examination of the environmental and social effects of these location decisions on the local area.

While the main aim is to examine the impact of industrialisation on Manchester, specific objectives are more multifarious. As far as content is concerned the objectives are:

(a) To impart knowledge of the range of early industries established in Manchester.
(b) The site requirements of those industries.
(c) The social and environmental conditions arising from the presence of those industries: 'externalities'.
(d) The perception and awareness of these conditions on the part of entrepreneurs establishing these industries.
(e) The perception of the conditions by local people.

Skills involved include map reading and the use of grid references and the social skills of co-operation in reaching a joint decision.

It is vital that the main aim of the exercise should be kept from the pupils at first. The exercise should be presented as a study of industrial location. Apart from what is written on their role cards, no mention should be made of pollution, health or hygiene although if a pupil decides to include pollution as one of the factors influencing his decision, this should of course be accepted. The exercise best fits into a course on industry and immediately after a lesson on industrial location.

Materials and teaching strategies

The following instructions for teachers and suggested timing are offered as a rough guide:

Instructions for teachers:	*Approximate timing*
1 Issue instructions sheets with map 2 Issue role cards at random to groups of two or three pupils	5 minutes
3 Groups discuss possible locations for their factories to make initial choices	10 minutes
4 Groups in turn indicate their chosen squares by calling out grid references for factory and houses. (An OHP master map is useful here and saves time.) Later groups may wish to change their chosen squares if these are chosen by earlier groups	20 minutes
End of first lesson if only singles are available	
5 On completion of map, issue sheet 2 and read these. Compare sites mentioned with those chosen by groups.	10 minutes
6 Issue sheet 3 which can be used either as basis of class discussion or as test	20 minutes
7 Draw conclusions	5 minutes

STUDENT SHEET ONE

The early industries of Manchester

Figure 1 is a map of Manchester as it was in the middle of the eighteenth century. It shows the rivers Irwell, Irk and Medlock which flow towards the southwest corner of the map to meet the River Mersey. The main roads are named after the towns and cities to which they lead. We are going to look at the decisions taken by the early industrialists of Manchester and the pattern of change they wrought on the landscape.

You will be given role cards each with a different industry marked on:

1 Cotton mill	6 Twine factory
2 Gas works	7 Glue factory
3 Bleach factory	8 Brewery
4 Dye works	9 Tannery
5 Shoddy factory	10 Rubber works

Each group will choose a site on the map for their factory or works in turn, bearing in mind the information on the role card and the fact that besides the square in which the factory stands you will have to build houses for workers within walking distance of the works (this means that you will have to build five squares of houses within three squares distance in any direction from the factory) for there is no public transport in Manchester at this time. Rivers can only be crossed by road bridge.

If another group chooses the square you chose before you can build your factory, choose an alternative site. The sites will be chosen in the order shown on the list above, so later groups should have alternative sites ready in case their first choice is used by others.

Make sure that other groups do not see your role card or obtain any information about your industry until you have sited your factory. When it is your turn your teacher will want to know the grid references of the squares you have chosen so *have these ready.*

Role cards

1 Cotton mill

Raw materials	Raw cotton, vast quantities of soft, lime-free water. This should be clean and as free from impurities as possible. Coal or coke important later as fuel for steam engines; this is often brought by river.
Site requirements	Access to water; mills are often astride streams for early power.
Market	National; much is exported. Cotton waste often passed on to the shoddy industry.

2 Gas works

Raw materials	Coal, water in large quantities.
Site requirements	Site needs to be close to running water. Large areas of flat land for buildings and storage of coal and gas.
Market	The gas is piped to all parts of the city so a fairly central location would seem useful.

The manufacture of gas in the nineteenth century was a highly noxious process. The area around the works was often covered in coal dust and the air was heavily charged with dust and fumes from the gas and from heavy chemical by-products such as coal tar. Hot waste water can be discharged into rivers.

3 Bleach factory

Raw materials	Caustic soda, chlorine, lime and sulphur; usually carried on water. Water used in process. Coke for heating.
Site requirements	Access to copious quantities of water. Flat land for building and storage.
Market	Mainly local cloth makers and cotton mills.

The acrid smell of the raw materials mingling with vast quantities of steam and smoke makes this industry a most unpleasant one. Often involved releasing hot chemically-charged water into the rivers.

4 Dye works

Raw materials	These vary according to the colour required but mainly consist of chemicals carried by water. Mauve dyes can be made from coal tar, a by-product from the gas works and other dyes require pigments; large quantities of water are used.
Site requirements	Large amounts of running water and flat land for buildings and storage space.
Market	Mainly local, particularly the cotton mills in and around Manchester.

Dyeing is a very smelly process mainly on account of the raw materials involved; waste products and filthy water is released steaming into the rivers.

5 Shoddy factory

Raw materials	Cotton rags, cotton waste and old clothes, rags, etc.
Site requirements	Large area of flat land for building and storage; access to running water.
Market	Local and national: many seamen's and workmen's garments are made from duffle together with stuffing for furniture and mattresses.

Much dust and fluff is released into the air besides smoke and steam; large quantities of dirty water released into the rivers from washing the rags. The storage of unwashed rags, often with vermin and the burning of oily rags which could not be used in the process is the most unpleasant part of this industry.

6 Twine factory

Raw materials	Indian Hemp, jute, bleaches and chemicals, tar and large amounts of water.
Site requirements	The works have unusual site requirements, apart from the usual ones of running water and flat land for building. Rope walks are long thin buildings, sometimes half a mile in length.
Market	Nationwide but a goodly proportion local especially packaging for factories.

Hemp and jute have a pungent smell and are stored on site; the smelling increases when the raw materials are rotted or soaked in water. Bleaching twine and tarring rope also is a strong smelling process.

7 Glue factory

Raw materials	Normally animal waste from slaughterhouses, fish meal from the fish markets (often rotten and unsaleable) and bones from the 'rag and bone men' sometimes stored on site.
Site requirements	Large amounts of running water and flat land for storage.
Market	Local packing factories, carpenters, joiners, etc.

The manufacture of glue is one of the most revolting sources of smells and pollution at this time. Apart from the smell and effluvia leaking from the piles of raw materials, the process itself involves heating the raw material, straining off the solid matter which was either burnt or thrown into the river. The oils in the glue itself have a penetrating smell and smoke from the furnaces hangs around the works.

8 Brewery

Raw materials	Hops and malt, sugar and yeast; large amounts of fresh water
Site requirements	Large amounts of running water, flat land for buildings and storage space, ideally a fairly central location for ease of distribution all over the city.
Market	Mainly local in and around Manchester

A brewery has a distinctive smell and smoke and steam are released into the air.

9 Tannery

Raw materials	New animal skins mostly imported via the River Mersey from the Argentine, South Africa, India and Ireland although many at this time will be brought by road from other parts of Britain. Oak bark from inland, tannic acid from chemical works further down the Mersey. Locally made oils and dyes, salt from Cheshire; large quantities of water.
Site requirements	Large amounts of running water for use and for transport; storage space and buildings will need flat land.
Markets	Mainly local for boots, and leather belting for driving mill machinery.

Tanning is a very noxious process with miscellaneous smells mingling; the stench of uncured skins combines with the caustic aroma of the chemicals involved, the sickening odour of burning waste and the pungent fumes of oils permeate the air. Waste is often tipped straight into the river or left in heaps on nearby vacant land.

10 Rubber works

Raw materials	Sheets of latex from Brazil, chemicals from the lower Mersey, coal and coke brought by water.
Site requirements	Access to running water, flat land for buildings and for storage. Water transport.
Market	Local and national for making industrial clothing, boots etc.

One of the later industries to be introduced into the area, vulcanisation is a process which inevitably produces a pungent and familiar smell; smoke and grime fill the air and waste products tipped into the river.

STUDENT SHEET TWO

From 'Human Documents of the Industrial Revolution'
The state of the towns: A gazetteer of disgusting places

The towns in which the people lived were, judged by modern standards, dreadful places, filled with repulsive sights and horrid smells, lacking in almost everything that makes for health and happiness ... The facts are beyond dispute: indeed it is almost impossible to exaggerate their abominable character.

In themselves the problems were not altogether new. There had always been overcrowding in the towns, where the houses were closely packed within the encircling walls and were unprovided with water supply or drains, when they were poorly lit and ventilated and the accommodation they provided was cramped and insufficient. There had always been a dunghill before the door which was not always wreathed in roses. Fever and other diseases were regular visitants and crime and vice were equally commonplace. Poverty was the common lot, and stench and squalor accepted as part of the natural order.

The Industrial Revolution, although it did not create the problems, made them infinitely worse. Those dunghills, for instance; in the old towns and villages, they were nothing like the menace they assumed when, unblown upon by country breezes, they made such horrid accumulations in the heart of the thickly populated city. Everyone was agreed about their nastiness; something ought to be done about them – but what?

We should not forget that we are living after more than 100 years of devoted and persistent attention to public health. The men of those changeful years, when the population was increasing by leaps and bounds and the rapidly growing towns were never big enough to take the numbers clamouring for house room, were very differently placed. They had nothing to go on. They simply did not know what to do.

There were no sanitary inspectors, no Environmental Health Officers, no statutary regulations to be complied with by order of the Council; there was no 'Council' until the Municipal Corporations were formed in 1835 and even then they took some years to get into their stride. Overcrowding was inevitable since there were hosts of people requiring to be housed as cheaply as possible, and since there was no mechanical means of transport, public or private, the dwellings had to be built within walking distance of the workplaces.

As the years passed and the problems became more pressing, a race of 'Sanitarians' arose, most, but not all, of whom were medical men. Their efforts were magnificent in the circumstances but they had such little knowledge to go on. Why, they did not even know that there were such things as germs! This explains the mention of noxious exhalations, 'effluvia' and something that was called 'miasma'....

Dr Kay's Manchester (from *The Moral and Physical Condition of the Working Poor in Manchester:* 1835):

'In some parts of the town exist evils so abominable as to require more minute description. A portion of low swampy ground, liable to be frequently inundated and to constant exhalation is included between a high bank over which the Oxford Road passes and a bend of the River Medlock where its course is impeded by weirs. This unhealthy spot, so low that the chimneys of its houses - some of them three storeys high, are little above the level of the road. About two hundred of these habitations are crowded together in an extremely narrow space and are inhabited by the lowest Irish. Most of these houses have also cellars, whose floor is scarcely elevated above the level of the water flowing into the Medlock ... these narrow abodes are in consequence always damp and on the slightest rise of the river are flooded to the depth of several inches which is a frequent occurrence. ... This district is surrounded on every side by some of the largest factories of the town, whose chimneys vomit forth dense clouds of smoke which 'hang in the air over this most salubrious region' ....

WHICH GROUP WAS RESPONSIBLE FOR BUILDING IN THIS AREA AND CREATING THESE CONDITIONS?

... 'The state of the houses between Store Street and Travers Street and the London Road is exceedingly wretched - especially those built on some irregular and broken mounds of clay on a steep declivity descending into Store Street. These narrow alleys are rough irregular gullies down which filthy water percolates and the inhabitants are crowded in delapidated abodes or obscure damp cellars in which it is impossible for the health to be preserved.'

... 'The Irk, black with the refuse of dye works erected on the banks, receives excrementitious matter from some sewers in this portion of the town - the drainage from the gas works and filth of the most pernicious matter from the bone works, tanneries, glue manufacturers etc. Immediately beneath Ducie Bridge, in a deep hollow between two high banks, it sweeps round a cluster of the most wretched and delapidated buildings of the town. The course of the river is here impeded by a weir; a large tannery, eight storeys high (three of which storeys are filled with skins exposed to the atmosphere in some stage of the processes to which they are subject) lowers close to this crazy labyrinth of buildings'...

A letter of Dr Robertson, a Manchester surgeon, 1840:

'The factories have necessarily sprung up along the watercourses which are the Rivers Irwell, Irk and Medlock and the Rochdale Canal. The dwellings of the work people have kept increasing on the confines of the factory districts. The interest and convenience of individual manufacturers and owners of property have determined the growth of the town and the manner of that growth, while the comfort, health or happiness of the inhabitants have not been considered ... Manchester has no Public Park or other ground where the population can walk and breathe the fresh air. In this respect Manchester is dreadfully defective, more so than any other town in the Empire. Every advantage of this nature has been sacrificed in the getting of ground rent...'

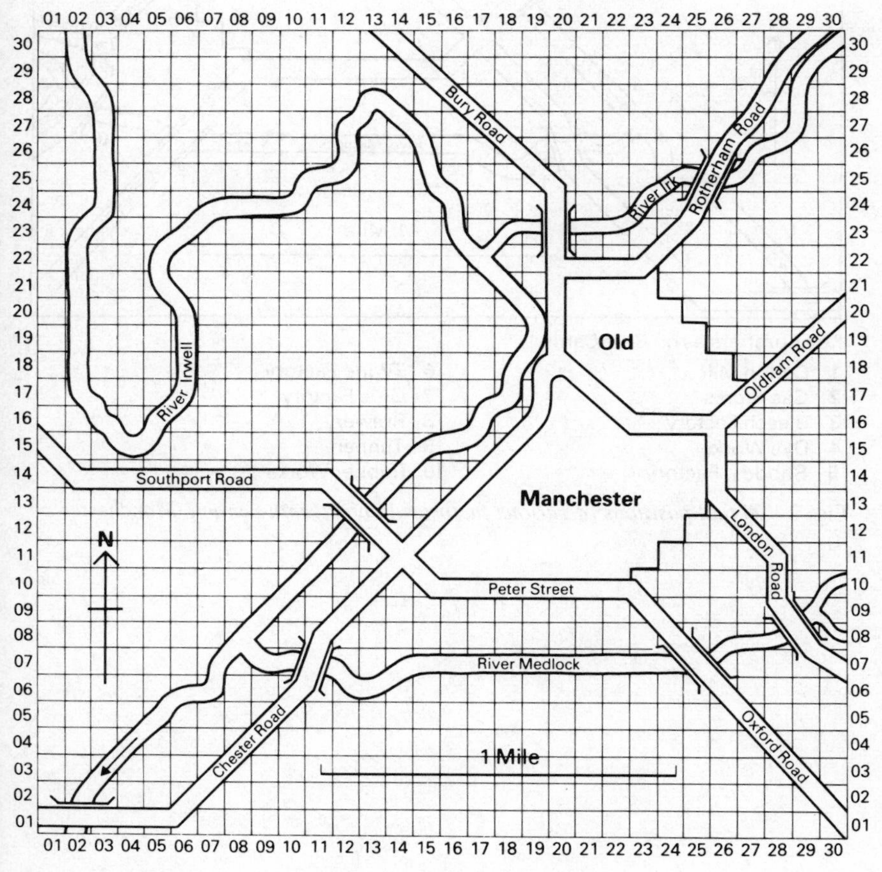

Fig. 1 *Manchester in the middle of the eighteenth century*

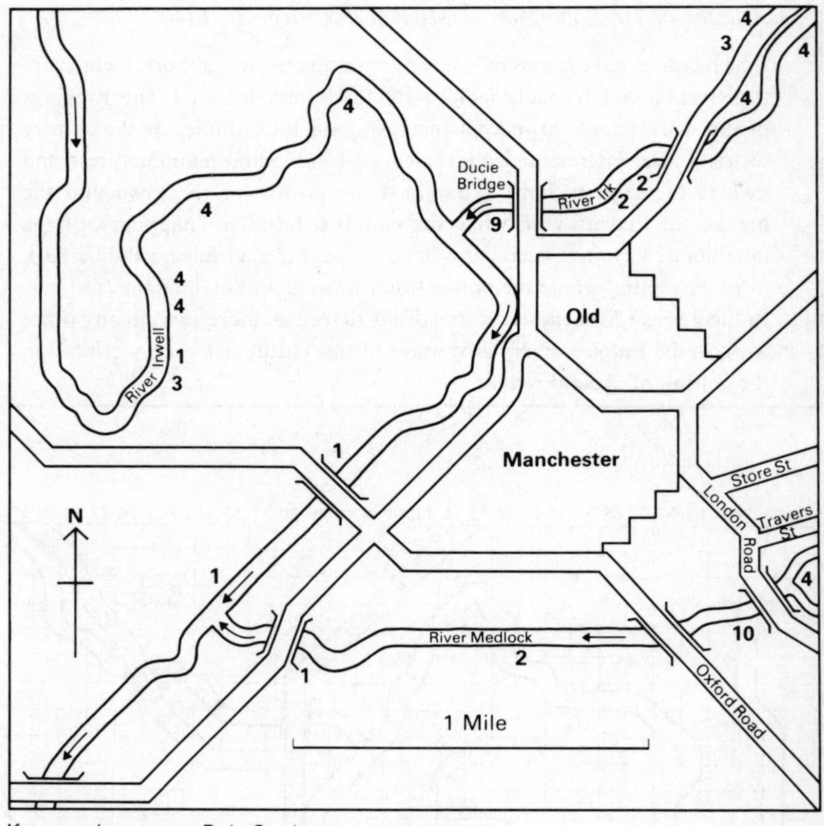

Key numbers as on Role Cards:

1 Cotton Mill
2 Gas Works
3 Bleach Factory
4 Dye Works
5 Shoddy Factory
6 Twine Factory
7 Glue Factory
8 Brewery
9 Tannery
10 Rubber Works

Fig. 2 *Actual positions of various factories in nineteenth-century Manchester*

STUDENT SHEET THREE

1 How many of the factories and works which you built coincided with those of the same type (a) mentioned in the extracts? (b) shown on Fig. 2?
2 What were the main site requirements of these early industries?
3 Why were the workers' houses built so close to the factories?
4 Draw a flow diagram to show how many of the factories described on the role cards are connected to each other by, for example, by-products from one factory being processed by another factory.
5 Do you suppose that the industrialists who built their works and factories were aware of the effects their factories would have on the area around? Did *you* realise when you built yours?
6 Make a list of the pollution effects caused by the works and factories and lack of sanitation described on Sheet Two. By each effect, write down the services provided by modern local authorities preventing these terrible conditions from existing today.

Further reading

Pupils wishing to read more about these conditions, their influence on writers, including Engels, who witnessed them, and about what was done about them could refer to the following:

Cootes, R. J. *Britain since 1700*. Longman Secondary Histories
Roberts, Martin, *Machines and Liberty: 1789–1914*. OUP

Further reading

BALE, J. (1978) 'Externality gradients', *Area* **10**(5)
HARVEY, D. (1973) *Social Justice and the City*. Arnold
SMITH, D. M. (1977) *Human Geography; a Welfare Approach*. Arnold

II Hunters and collectors – *a study in changing pupil perceptions of other societies*

John Bale

The behavioural geography which has emerged in the last decade has drawn attention not only to the perceptions of decision-makers but to the stereotypical myths which colour most people's view of the world in which they live. To some extent, we are all the prisoners of our own 'private geographies', and the old 'objective' world once so cherished, as the gift which geography teachers could bring to the education of the young, now seems properly categorised as mirage.

But the creation of a 'global village' through improved telecommunication has no doubt increased the amount of world knowledge which everyone has, and the subject-matter of geography in many classrooms is now the negotiation of changing such views as are *already* held. (It is to be hoped that the teacher may have more evidence than his or her students for such views, but there is no guarantee of this in an age when globe-trotting parents take their offspring for package holidays to Japan, Sri Lanka or Brazil.)

John Bale's unit about 'primitive peoples' was deliberately designed to take some of the most hallowed basic material of geography and reveal its Eurocentredness. He takes evidence about Aborigines and Red Indians and uses a profiling exercise before and after its use to see what stereotypes exist and what can be changed.

'In the words of one best-selling geography text-book, these people live in 'regions of difficulty'. But difficult for whom?'

The material has been used with classes of 13 to 15-year-olds, but can also be readily adapted for oral use with much younger groups.

Introduction

Hunters and collectors – the most primitive peoples, according to most geography texts – lived in a society of scarce wants and unlimited resources. Stone-age economies provided a milieu in which a smaller proportion of the population than at present went to sleep hungry; the 'primitive' aborigine had more leisure-time than present-day Westerners, worked fewer hours per week, consumed a more balanced diet and underutilised the environment.

We tend to think of hunters and collectors as poor because they had few material possessions; perhaps it would be better to call them free. Poverty seems to have grown with 'civilisation'. Indeed, many of the hunters and collectors were apparently indifferent to material possessions. Who then has the higher social and moral attainment, stone-age hunters or modern suburbanites? As Brookfield[1] has reminded us, many Western indicators of *progress* are, in fact, indices of degradation, seen from the African perspective.

Materials and teaching strategies

The materials consist of three work-sheets

The purpose of Assignment I is to indicate to the students that hunters and collectors possessed a lifestyle not only different from twentieth-century society, but different also from the biased viewpoints presented in traditional texts. The student has to analyse the graphs, and is likely to conclude that work was not hard for the aborigines of Arnhem Land; their leisure time exceeded that enjoyed by most people of the twentieth-century Western world.

The assignment might be used as homework, following an initial class discussion of views about such peoples.

The purpose of Assignment II is to use a technique known as 'semantic differential profiling' to elicit from children their images of native Americans, following an examination of some native American settlement patterns.

Experience suggests that the term 'red indian' usually possesses certain connotations for some students and that these are often unfavourable. Figure 3 demonstrates this in relation to a group of twenty-seven 14-year-old children of average ability from a school in a working-class area of North Staffordshire.

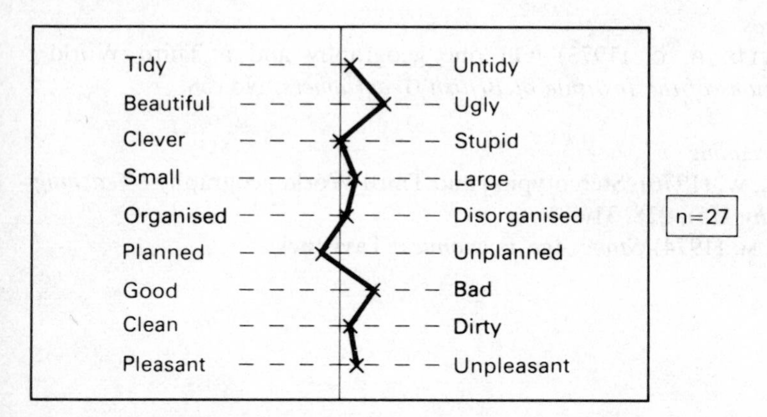

Fig. 3 *Semantic differential profile for concept 'Red Indian'. The profile is constructed by obtaining the average score for each scale of bipolar adjectives.* (*See Assignment II*)

Students may arrive at more positive perceptions, however, if they have the chance to study maps such as those shown in Assignment II.

An example of a semantic differential profile for students exposed to such maps as those in Assignment II is shown in Fig. 4.

In this case, the average semantic differential scores, as expressed in the profile, reveal a more favourable attitude than was the case when 'red indians' was provided as a stimulus term without any other knowledge.

It is suggested that this type of material might be used in classrooms where it is the teachers' intention to provide a sympathetic view of unfamiliar cultures and peoples.

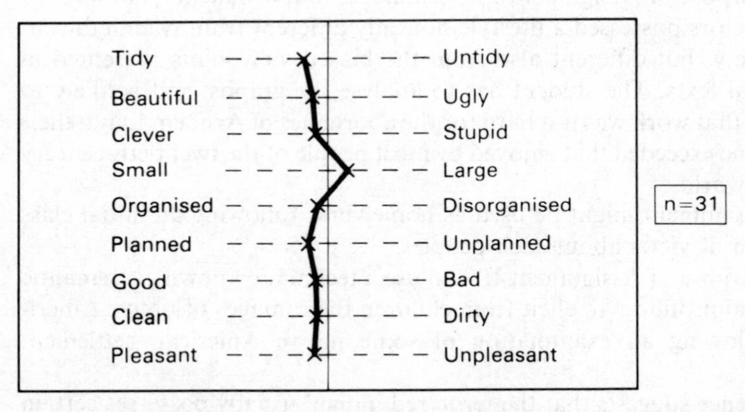

Fig. 4 *Semantic differential profile based on reactions of a sample of thirty-one secondary school students to the diagram shown in Assignment II*

References

BROOKFIELD, H. C. (1973) 'On one geography and a Third World', *Transactions of the Institute of British Geographers*, No. 58

Further reading

MARSDEN, W. (1976) 'Stereotyping and Third World geography', *Teaching Geography* **1**(5), 228–31.
SAHLINS, M. (1974) *Stone Age Economies*. Tavistock.

ASSIGNMENT I

Homework assignment.

The Aborigines of Fish Creek, Arnhem Land, Australia

The graph opposite shows the average number of hours worked per day by the aborigine men in part of Arnhem Land. The figures for women are similar. The graph indicates time spent at work, doing things like hunting, plant collecting, preparing foods, and repairing weapons.

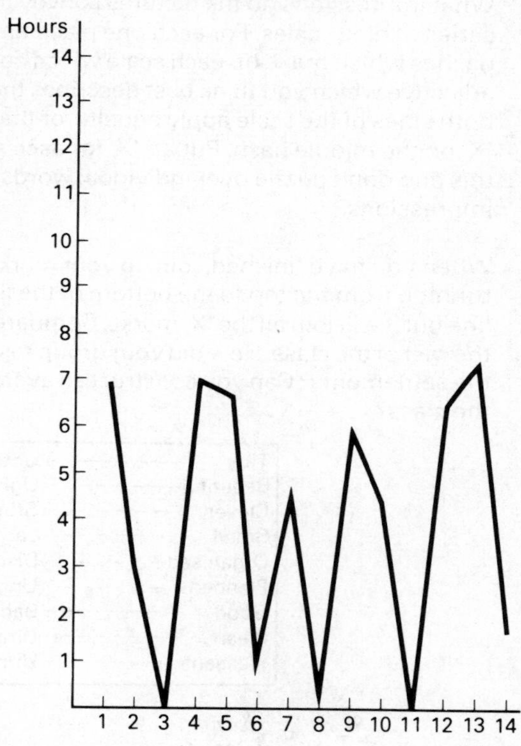

Tasks:

1 Add to the diagram, another graph showing the number of hours worked each day by an adult in your family over a period of 14 days.
2 What was the mean number of hours worked per day for the aborigine men?
3 What was the median?
4 What was the mode?
5 What was the average number of hours worked per day by the adult from your family?
6 What was the median and mode?
7 Try and find out the number of hours people worked in working class Victorian times. Add the results to the graph.
8 Who spent more time working and resting, the aborigine or modern man?
9 Can you suggest reasons why, for your answer to q.8.

(Source of original data: M. Sahlins, 1974, 15)

ASSIGNMENT II

Study the following maps of native American settlements

What impressions do the pictures convey to you? Given below are a series of nine scales. For each one place an 'X' on one of the seven dashes which make up each scale which comes nearest to the adjective which you think best describes the settlements. If you think both sides of the scale apply equally, or that neither apply, place your 'X' on the middle dash. Put an 'X' for each scale. Work quickly through this and don't puzzle over individual words. Give your first impressions.

When you have finished, join up your marks so that you get a line running from the top to the bottom of the list. It may not be a straight line but it will join all the 'X' marks. Compare your profile with those of the rest of the class. How did your group respond to the appearance of the settlements? Can you construct an average profile for the whole of the class?

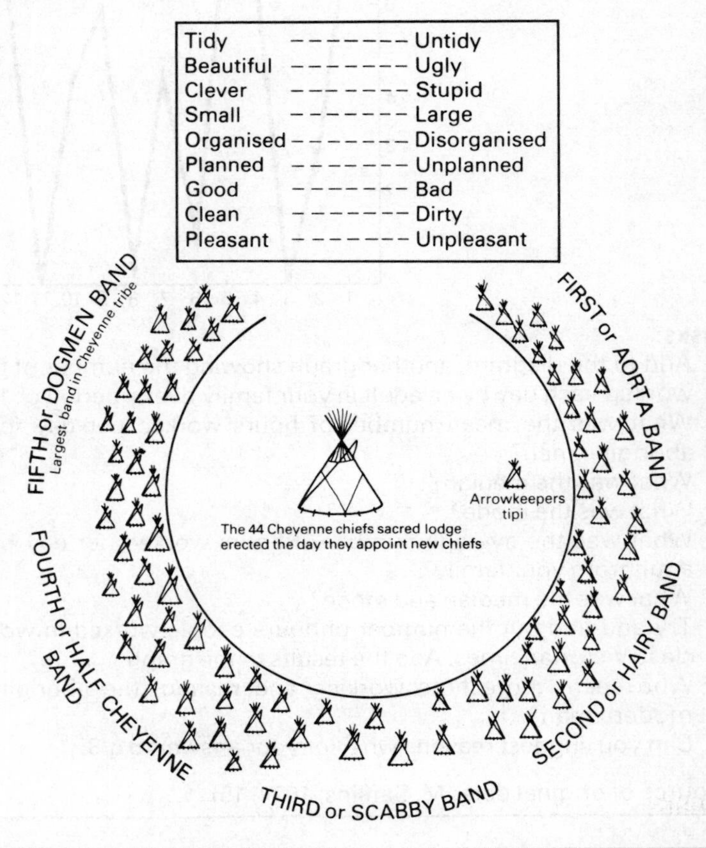

Tidy	– – – – – – –	Untidy
Beautiful	– – – – – – –	Ugly
Clever	– – – – – – –	Stupid
Small	– – – – – – –	Large
Organised	– – – – – – –	Disorganised
Planned	– – – – – – –	Unplanned
Good	– – – – – – –	Bad
Clean	– – – – – – –	Dirty
Pleasant	– – – – – – –	Unpleasant

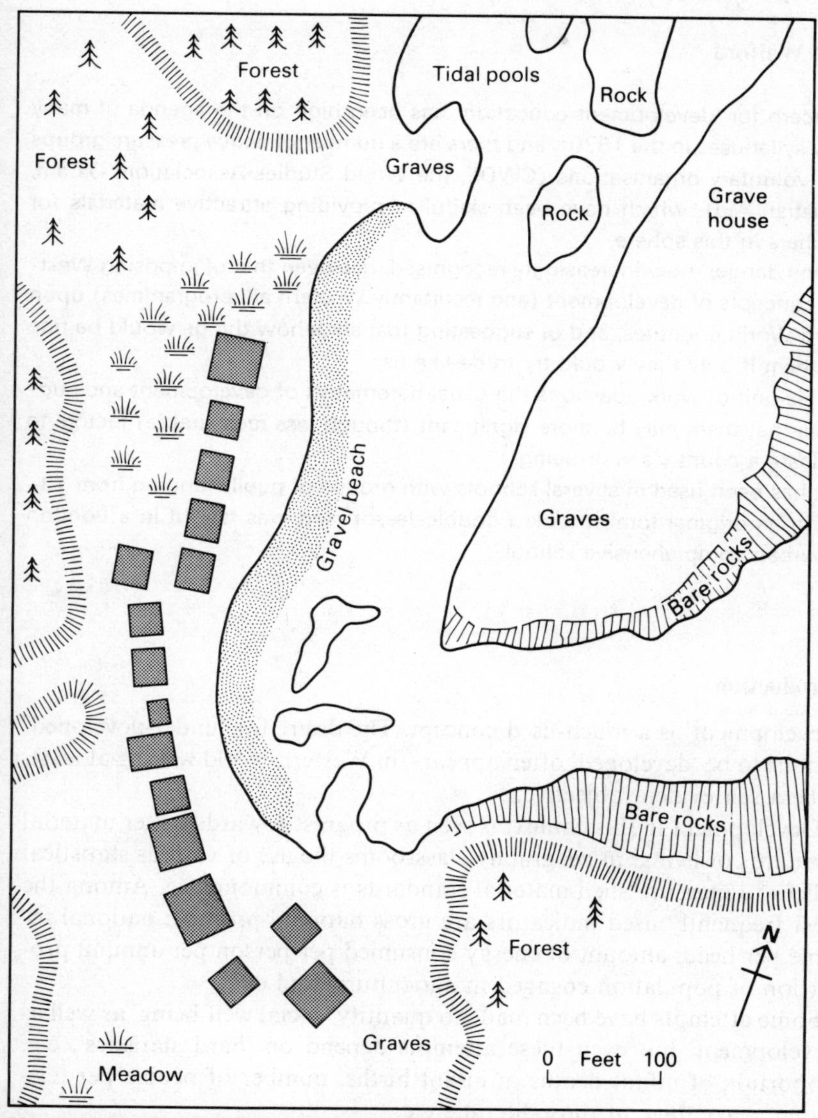

Plan of Ninstints village, Queen Charlotte Islands. British Columbia, Canada.

(Source of map: D. Fraser, *Village Planning in the Primitive World,* London, Studio Vista, n.d.).

III Does development make you happier? – *an exercise in quantifying the qualitative*

Rex Walford

Concern for 'development education' has been high on the agenda of many new syllabuses in the 1970s, and there are a number of active pressure groups and voluntary organisations (CWDE, the World Studies Association, Oxfam, Christian Aid), which have been skilful in providing attractive materials for teachers in this sphere.

One danger, now increasingly recognised, has been that of imposing Western concepts of development (and resultantly Western aid programmes) upon Third World countries, and of suggesting that somehow things would be fine for them if only they would try to be like us.

This unit of work questions the usual barometers of development and suggests that there may be more significant (though less measurable) factors to indicate a country's well-being.

It has been used in several schools with groups of pupils ranging from 13–17; in its original form it took a double-lesson and was taught in a London suburban comprehensive school.

Introduction

'Development' is a much-used concept. The desire for 'under-developed' nations to be 'developed' often appears, in Western-world writing at least, to be a self-evident objective.

Development in this context is seen as progress towards better material standards of living; in geography classrooms the use of various statistical indices to measure such material standards is commonplace. Among the most frequently used indicators are gross national product; national income per head; amount of energy consumed per person per annum; proportion of population engaged in agriculture; and etc.

Some attempts have been made to quantify 'social well-being' as well as 'development' but even these attempts depend on 'hard statistics', e.g. proportion of infant deaths in infant births, number of people per telephone, percentage of those literate, etc.

Few attempts have been made to quantify 'happiness' since it is difficult to find any 'hard data' which could be used. However, this is surely a poor reason for *ignoring* happiness as a possible goal towards which some societies may strive. It may be that the accretion of wealth to produce material prosperity and development does not go hand in hand with increasing happiness; Biblical precept suggests rather the opposite.

An article in *The Times* provided the germ of the idea for this lesson, and it provided the following thought. 'It is clear that Britain is "developed" and Swaziland is "under-developed" (or euphemistically "developing"); but is Britain happier than Swaziland?'

Materials and teaching strategies

The unit, as described here, took up a double lesson of 80 minutes and a homework period. Beforehand, the teacher gathered background material (slides, books, newspaper articles, material from voluntary agencies) about Swaziland; he also had the benefit of some personal experience; clearly, the latter is a key factor in determining the country chosen for comparison in this way, since it is an added bonus. It is not essential, however.

1 The lesson began with a general discussion about 'development' and the teacher injected into the discussion the question 'Does greater prosperity necessarily make you happier?' There was disagreement about this, with the class roughly dividing into (a) materialists ('Of course, you're happier if you're rich'), (b) the hair-shirt brigade ('Money doesn't matter – what matters is helping others') and (c) those who didn't seem to know or care.

2 After 10 minutes, the teacher began to seek a definition of happiness. Using the blackboard, and aided by class suggestions, he built up the following composite definition:

Happiness = Security = being part of a caring community (family/friends)
= being 'at home' in the environment in which you live
= Satisfaction = fulfilment of physical needs (food, drink, sleep)
sexual drives
creative drives (leisure activities, sports)
aesthetic needs (music, art, dance)

The teacher had the plan in his mind before starting and 'interpreted' pupil contributions into these categories; there was no dissent about this. The question of 'sexual drives' was recognised as a delicate issue, but it was raised by the pupils themselves in a slightly self-conscious and jocular way – the teacher took the contribution with seriousness and the jocularity soon subsided when it was seen that the matter was not considered taboo.

3 The teacher led a discussion about which of these six aspects of happiness was most important. Pupils made their own private assess-

ments on a piece of paper and results were compared. As a result of the class's views, the following weighting system was drawn up:

Being part of a caring community	20
Being 'at home' in the environment	15
Satisfaction of physical needs	35
Satisfaction of sexual drive	10
Satisfaction of creative drives	15
Satisfaction of aesthetic needs	5
	100

4 The class was then shown some slides of Swaziland, and the teacher talked about his experiences there. They were also given some written extracts of material and some photographs. It emerged that there was little actual starvation in Swaziland (though not always good nutrition), that the social support of extended families was strong, that the Swazis were not 'at odds' with their environment, that there were chances for leisure and recreation, and that Swazi scenery and art had beauty and vigour.

5 A briefer discussion considered the same elements in relation to the United Kingdom, with the class being encouraged to draw on their own experiences. Some spoke vehemently about faceless urban environments and the lack of things to do in their own neighbourhoods.

6 The class were set the task of assigning points (out of 100) for each category to both Britain and Swaziland, based on 'the life of the average citizen in each country'. They were grouped into twos or fours, as they wished; heated arguments amongst the groups ensued as they tackled the task. The points score task lasted for 15 minutes.

7 The group scores were gathered publically and compared. Every group rated Britain higher for 'satisfaction of physical needs' by a considerable amount; the countries were rated equal for satisfaction of sexual drives; in the other categories, overall Swaziland was rated higher.

8 The teacher built up the following table from the average of the group scores, and those with calculators worked out the category score × the weighting, and also the final totals. The table looked like this:

		Britain		*Swaziland*	
Category	*Weighting*	*Score*	*W × S=*	*Score*	*W × S=*
Being part of a caring community	20	40	800	80	1 600
Being at home in the environment	15	50	750	70	1 050
Satisfaction of physical needs	35	95	3 465	70	2 450
Satisfaction of sexual drives	10	75	750	75	750
Satisfaction of creative drives	15	30	450	60	900
Satisfaction of aesthetic needs	5	40	200	60	300
Totals			6 415		7 050

9 There was some discussion of the conclusion that the tables reached. The teacher cast doubt on the *accuracy* of the tables and said it could only be a rough guide. 'But if development is the *same* as happiness,' said the teacher to close, provocatively, 'perhaps we had better start thinking of Britain as under-developed.'

Conclusions

The actual figures are pure conjecture, but the attempt to put some quantification on to qualitative ideas helps to explore the issue rather than dismiss it. It would have been possible to revise the weightings, and perhaps to be more favourable to Britain in some of the judgements; but the point is effectively made.

One obvious danger of this approach is the possible conclusion that it might be just as well therefore to leave countries like Swaziland without any aid at all. This becomes a rather grisly parody of the 'leave the natives happy in their mud huts' argument.

The discussion must be taken further to consider what kind of aid can sensitively be distributed to Third World countries without tearing apart the strengths of their individual and social lives; it leads to the paradox of development and aid in its deepest form.

IV Murderer at large! – *a simple map exercise that introduces the computer*

Peter Fox

Chilwell Comprehensive School, where Peter Fox teaches, is a purpose-built 11–18 comprehensive school on the outskirts of Nottingham with about 1 700 pupils.

The geography department has had a hand in the development and use of computer-assisted learning, since a terminal was loaned from Trent Polytechnic some years ago. The school now owns its own micro-processor and so is independent of outside links.

Chilwell is one of an increasing number of schools who use programmes useful to geography teaching made available by the Geographical Association Package Exchange (GAPE), administered by David Walker from the Geography Department at the University of Loughborough. (See Ch 10 in Part Three of this volume.)

Chilwell's adaptation of a program known as Hunt the Hurkle is described here in a unit of work which is concerned with developing and improving mapping skills, of mixed-ability Second-year pupils.

Other uses of the computer at Chilwell include helping students with individual field project data (using programmes such as TRAFFIC, CORR, and GRAVITY) and developing concepts of industrial location with O level and CSE classes in the fifth year).

Introduction

Our association with computer-assisted learning has not been without its difficulties - at first only a limited user time available, remote access, a large telephone bill, and a need to learn the computer's language (BASIC). But the purchase of a micro-computer for the school and the purchase of transferable program packages has eased the strain.

As with any teaching unit, the teacher needs to ask if the program is suitable to the age and ability of the class before using it, and is the lesson relevant to the syllabus. Careful lesson preparation is important; the program needs to be well pupil-proofed so that things do not go wrong with the software when it is being used by individual pupils or groups.

If pupils have not used a computer before they may tend to see it as a toy or TV game and be mesmerised by it; such novelty value soon passes with repeated use.

A more practical problem may be finding enough plugs to provide outlets of power for the various pieces of equipment, e.g. the teletypes, the visual display units (TV monitors) and the processor itself.

Our Murderer at Large game was used with second years to:

(a) introduce the computer as an educational tool;
(b) test the understanding of the relationship between grid references and compass directions;
(c) reinforce the idea of grid reference deduction;
(d) motivate the pupils in the use and discovery of grid references;
(e) improve other map-work skills.

Materials and teaching strategies

1 Students are provided with copies of an OS 1:50 000 map or an extract. (Ideally, it is best to use a sheet which has a large number of roads, and woodland symbols.) They are also given tracing paper. The teacher needs access to the micro-computer, a large visual display unit and a computer program. (The Hunt the Hurkle program devised by I. D. H. Shepherd and available from the Geographical Association Package Exchange, c/o University of Loughborough, is a suitable program for this activity.)

2 The teacher tells the class that a murderer is said to be at large in a

10 × 10 km square on the map; they are to play the role of the police and try to catch the murderer.

3 The teacher then delimits the area where the murderer is to be found by drawing in area chosen from the map on the blackboard. Pupils draw or trace their own 10 × 10 km grid (20 cm × 20 cm on to a 1:50 000 map).
4 The teacher recaps (or teaches if necessary) how to identify and write four-figure grid references, and compass directions; students construct a simple eight-point direction-rose for their own use.
5 One student is selected to come to the micro-computer, to type in a number to activate the program and to select four-figure references as the level of use of the program.
6 The teacher asks the class to guess where the murderer might be or where it might be best to start the search. (The best place to start is the middle of the area, and students using the game several times come to understand this).
7 A starting place for the search is chosen, and typed into the computer. Students mark this on their own grids.
8 The computer then responds with a 'tip-off' as to the location of the hiding murderer (e.g. 'Go south-west').
9 Students now check their own maps and delimit the area where they think the murderer may be found. They then 'block all roads and footpaths' in pencil on the tracing paper over the map.
10 Another student is asked to put a guess into the computer based on this evidence, the computer will again respond, 'Go west'.
11 Students are again asked to delimit this new smaller area and move in on the murderer.
12 The exercise is repeated until the students hunt down the murderer's hiding-place; the computer will then print CONGRATULATIONS! on the screen. But only the square where the murderer is hiding has been found. Students are now asked to examine the grid square and make a list of suitable hideouts from the OS map itself.
13 The disadvantages of the vagueness of four-figure grid references are thus made apparent, and the exercise can be used again with the computer generating a different hiding-place and responding to six-figure references.
14 The class can be split into groups and each one asked to plan a search strategy, following a discussion of the features of the area as shown on the map. (This can lead to consideration of other types of searches, e.g. for oil/coal/minerals/new homes in strange towns, and allow groups and individuals to work with the computer subsequently, using their own ideas.

Problems encountered

Students who have not seen a computer before waste a great deal of time if they misunderstand instructions for use.

Students are sometimes liable to get co-ordinates the wrong way round.

Less-able pupils need some help in interpreting the map itself for hiding places.

Able students often question the wisdom of a murderer if the computer should happen to locate him in a square with few hiding places!

V Changing urban geography – *using house prices as a data source*

Andrew Kirby

Andrew Kirby's exercise is in the practical quantitative tradition which has infused much new geographical work in schools during the 1970s. It uses readily available data sources to create a graph and a regression line; residual points are then identified and they in turn provide a focus of interest, since they identify irregularities of a spatial pattern.

But it is significant that the data content – house prices and rateable values – relates to rather less traditional areas of urban study. It reflects a concern with housing markets, movements of capital, ownership factors and political power as explanations of some of the urban patterns found in the contemporary world. It shifts the focus away from some of the simple determinants which the first generation of urban models seemed to propose.

Introduction

In their work *Inside the City*, Everson and FitzGerald[1] synthesised the bulk of the material that a decade ago constituted urban geography. They introduced numerous concepts and models, and provided outlines of how these ideas could be tested in a local context.

In the intervening years, two things have changed. First, the world has changed – in other words the cities themselves have been transformed. Second, the ways in which urban geography approaches its subject have also changed: the things that it emphasises, and even the ways that research is undertaken, have altered.

This is hardly the place to examine in detail the changes that have occurred to our cities over the last ten or fifteen years. We can however summarise these (see Table 2):

Table 2 *Urban change in Britain*

Aspects of change	*Implications*
Regional performance	Old metropolitan centres declining: overspill and New Towns growing. Urban hierarchy changing.
Population out-movements	Urban cores losing population to suburban areas.
Transportation	Increasing car-ownership, greater mobility.
Ethnic change	Greater segregation of population within cities.

The list in the table is in no sense exhaustive: it does however reveal some of the ways in which our urban areas are evolving.

Clearly, when faced with this 'new' animal, urban geography has had to revise its opinions, alter its assumptions and even refocus its interests. One of the implications of the urban changes that have occurred is that there now exists a great deal of polarity within towns and cities: put more simply, poverty is more starkly revealed than before. This has caused geographers to focus upon its manifestations, and to account for both its spatial occurrence (in for example the inner city) and its social causes.

The net result of this transformation is that there now exists a great deal of research activity in relation to urban issues. Moreover, because 'new' questions are being asked – such as, for example, where do building societies lend their funds? – the emphasis is very much upon investigation, rather than the attempt to justify the use of a particular model. (Another implication of urban change is that the urban models generally discussed in undergraduate and sixth-form texts are even more inapplicable than was the case before.)

An excellent example of this type of research activity is revealed in a recent work by John Short[2] (1980), which documents several of these current interests, and, more important, the ways in which they can be replicated, using local data sources. In particular, Short focuses upon the residential structure of the city, the ways in which people move within the city and the ways in which housing has a role to play in these processes.

The following exercise is a simple example of this type of approach, using house prices and rateable values to find out something of a local housing market.

Materials and teaching strategies

The study relates for simplicity to a rural area, although the general principle is designed to be replicated in the urban context. The basic requirement is simply a collection of estate agents' prospectuses, which contain (a) house prices and (b) rateable values. (The author does not of course claim originality in the use of these data sources: see Everson and FitzGerald[1], pp. 24–5, and 128–9.) The data used in this exercise are reproduced in Table 3.

From the table, it can be seen that 32 dwellings have been included; 10 each in Wallingford and Didcot (which are both towns), and 6 each in the villages of Benson and Cholsey. For each dwelling, we possess both the price and the Rateable Value (RV). The *price* of a house can be regarded as a measure of its desirability, which in turn will reflect its size and its location. Similarly, the RV is an objective, valuer's estimate of the dwelling's size and external characteristics (including the neighbourhood). When we compare the two values for each property, we find a relatively close correlation ($r = 0.6$); the degree of association is also illustrated in Fig. 5, which shows a bivariate plot of price and RV.

Of course, this relationship is not perfect, and some properties command higher – or lower – prices than the RV suggests. We can illustrate this by predicting from the RV the likely price of any particular house; we

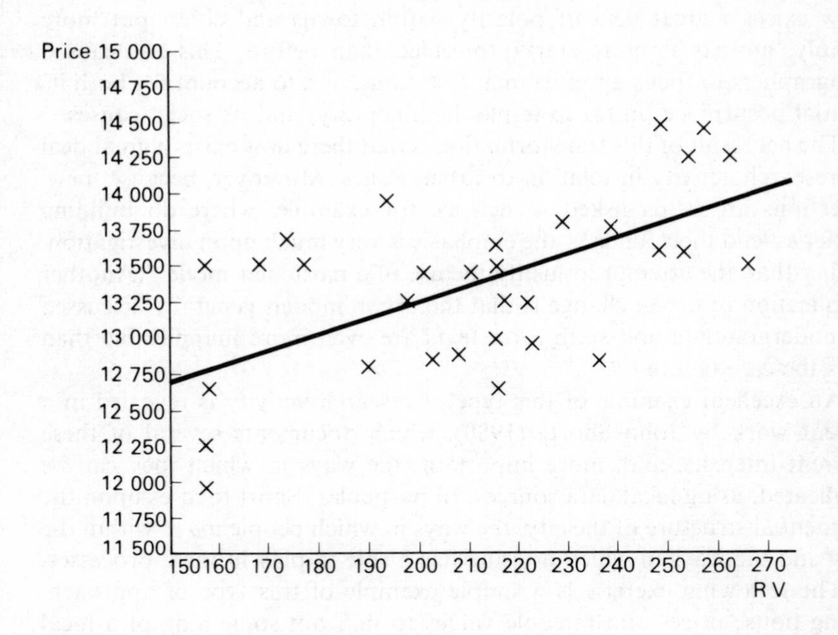

Fig. 5 *Bivariate plot of price and RV, data in £s*

do this by using the regression equation given in Table 3, which has been derived from the price/rateable value relationship for all 32 cases. These predicted prices are shown in Fig. 5 as the straight line on the graph. For any particular property, we can measure the differences between the predicted price (based on its RV), and the actual price: this difference is known as the *residual*.

As Fig. 5 and Table 3 show, there is a random pattern of residuals - or is there? One clear finding is that the village of Benson is *desirable*: people seem prepared to pay far more for housing there than we would expect given the rateable values (five of the six residuals are positive). Conversely, Didcot is *less desirable*: six of the ten residuals are negative. From this simple beginning, we can then progress to some further analysis to account for this (in this instance, Benson is a pleasant village, close to large research centres which offer good employment, whilst Didcot is a railway junction with a large power station close to the town).

This type of analysis can be transferred fairly simply to data collected by neighbourhoods within urban areas. Regression equations can be arrived at using either school computers or by hand, applying the formula outlined by Gregory in *Statistical Methods and the Geographer*.[3] Analysis of residuals can be undertaken to identify the residential structure of the local area, with this perhaps in turn providing the basis for further investigation: residential desirability can be compared with the distribution of attractive phenomena (like open space: Short,[2] p. 141) or undesirable phenomena (like football grounds): see Bale (Unit X).

Summary

The purpose of this kind of exercise is two-fold. First, and most important, it has as its aim the investigation of urban issues. Using tools (like house prices), we can find out a great deal about our cities: not a static measurement of land-uses, but rather as part of a more involved analysis of social issues. Second, we can use this type of study as a practical means of applying numerical methods. Quantification should not be seen as a mere task of learning: instead it must be viewed as a practical skill, like map-reading. Correlation and regression exercises are relatively simple ways of applying these skills, whilst also investigating topics in a fairly rigorous manner, which must ultimately be the twin purposes of any fieldwork.

References

1 EVERSON, J. A. and FITZGERALD, B. P. (1972) *Inside the City*. Longman
2 SHORT, J. R. (1980) *Urban Data Sources*. Butterworth
3 GREGORY, S. (1968) *Statistical Methods and the Geographer*. Longman

Further reading

KIRBY, A. M. (1981) 'A contemporary approach to urban areas', *Teaching Geography*, **6**(3)

Table 3: *Data for price/RV regressions*

		House price (X_1)	*Rateable value* (X_2)	*Predicted price*	*Residual*
	1	13 650	215	13 420	229
	2	13 250	217	13 442	−192
	3	13 000	215	13 420	−420
	4	13 700	174	12 964	735
	5	13 750	238	13 675	74
(Wallingford)	6	13 500	215	13 420	79
1–10	7	14 250	254	13 853	396
	8	13 500	266	13 986	−486
	9	12 250	157	12 776	−526
	10	13 600	248	13 786	−186
	11	13 450	201	13 264	185
	12	12 850	236	13 653	−803
	13	13 450	210	13 364	85
	14	12 650	215	13 420	−770
(Didcot)	15	12 800	190	13 142	−342
11–20	16	12 850	202	13 275	−425
	17	11 950	157	12 776	−826
	18	12 875	207	13 331	−456
	19	14 450	257	13 886	563
	20	13 950	194	13 186	763
	21	13 500	177	12 998	501
	22	13 500	168	12 898	601
(Benson)	23	13 250	198	13 231	18
21–26	24	14 250	254	13 853	396
	25	13 500	157	12 776	723
	26	12 650	158	12 787	−137
	27	14 500	249	13 797	702
	28	14 250	262	13 941	308
(Cholsey)	29	12 955	222	13 497	−542
27–32	30	13 600	253	13 841	−241
	31	13 250	222	13 497	−247
	32	13 800	228	13 564	235

Regression equation: $X_1 = 11\,033 + 11.1\ X_2$
Correlation coefficient: 0.6

VI Fourth year urban fieldwork – *hypothesis testing in Cheltenham*

Derek Plumb

Cleeve School in Cheltenham is an 11–18 comprehensive school with 1 550 pupils. Approximately 200 pupils usually opt for geography at the end of the third year, and Derek Plumb, as Head of Department, endeavours to make sure that all fourth and fifth year pupils are involved in fieldwork exercises during their progress towards either O level (Schools Council 14–18 Project examination) or a parallel Mode 3 CSE syllabus.

The unit described below is used as a coursework assessment in the course, under the category of 'Innovation Studies' which the examination allows. The aim here is to allow pupils to develop enquiry skills in geography, as well as to allow individual members of staff to experiment with new material and teaching approaches.

This fieldwork has, at its core, a day when 200 pupils are purposefully at work en masse in an urban environment – no mean feat!

Introduction

The Cheltenham urban field study follows a settlement topic which lasts for approximately eight weeks and it is a sequel to a three-week section of the course on the 'Internal structure of cities'. The topic for fieldwork study is narrowed down to a consideration of the urban patterns which exist within and near commercial areas.

The lesson plans for the unit in which the fieldwork occurs are summarised in Table 4. The overall aim of the unit is to show that all settlements have a recognisable internal division of morphology and function, and that these can change over time.

Concepts emphasised are *location*, *process* and *time*.

Pupils are encouraged to identify the processes operating in towns and cities and not to see urban models as static. Field observation quickly reveals the distortions of such models in reality.

The procedure outlined in Fig. 6 has the advantage of allowing pupils to reflect on why they are doing fieldwork and encourages them to identify and explain distortions to expected patterns.

We ask pupils to test two of the following four hypotheses in their fieldwork:

1 That the level of services can be used to identify different parts of a town.
2 That main roads influence the morphology and functions of buildings which have frontage to them.
3 That the greatest amount of renewal of buildings is in the central area of a town.
4 That there is a 'zone of blight' around the central area.

Table 4 *The internal structure of cities. A summary of the unit*

Teaching objectives	*Teaching theme, method*	*Time allocation*
To illustrate that:		
1. The land use of urban areas can be divided into major categories. Different zones of land use can readily be identified	Analysis of land uses with a town. Construction of graphs showing percentages of land use categories in different parts of a town.	1 double lesson 1 homework
2. Land values affect the nature and intensity of land use within urban areas 3. Lines of transportation affect the nature and intensity of land use within urban areas	Elementary land use theory. A study of processes involved in urban models. Burgess, Hoyt, Harris–Ullman and Mann	2 double lessons 2 homeworks
4. Functions group or disperse according to their specific needs	Analysis of the C.B.D., the core and the fringe.	2 double lessons 2 homeworks
5. Residential areas of cities vary in their design, social and ethnic make-up	Analysis of residential patterns. Plotting and graphing of ward data of selected cities	1 double lesson 1 homework
The teaching objectives are tested in the field study exercise	Fieldstudy, Cheltenham Processing fieldwork data Testing the hypotheses	1 morning 2 double lessons 4 homeworks (2 weeks)

Materials and teaching strategies

Before setting-off, a double lesson is devoted to a discussion of the significance of the hypotheses and to methods of recording and processing fieldwork data. The standardisation of 'what' and 'how' to record information

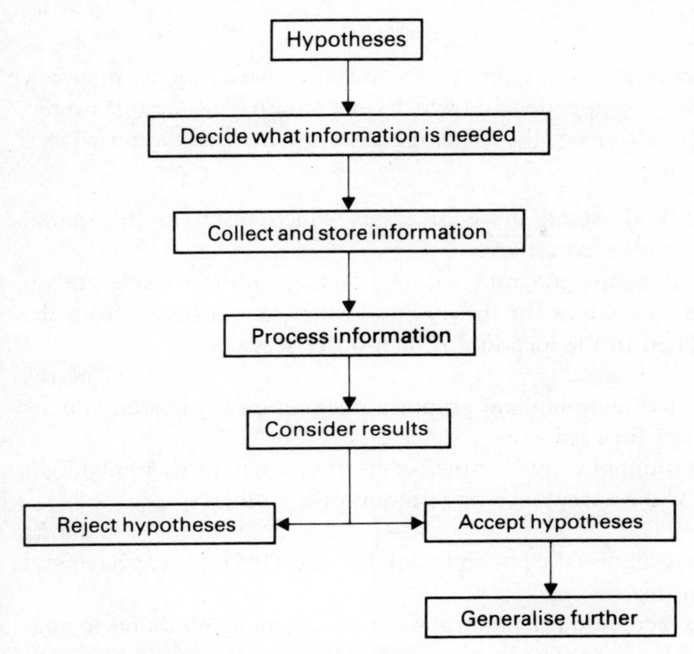

Fig. 6 *Hypotheses testing in fieldwork*

has proved vital if use of the fieldwork data is to be made subsequently in class. It is also equally important for the pupils to feel confident in what they are doing.

Pupils record the fieldwork data in their notebooks, noting buildings separately and classifying them by function (using a standard classification list), number of storeys, age (using a four-fold classification), condition (using a four-fold classification) and amount of renewal (using a three-fold classification).

Before the day itself, buses are booked, pupils insured and the local police informed of the operation. Then comes the day when 200 children armed with clip-boards descend on Cheltenham! The central area of the town is divided into four sections, and children are allocated sectors to cover: a 1:25 000 base map for each pair is provided to assist this.

Back in the classroom, two double lessons and two homeworks are devoted to the organisation of maps and graphs from the raw data. Pupils decide on the hypotheses in which they are most interested and then become responsible for collecting and mapping the raw data connected to these. Another four homeworks are allocated to the pupils' individual interpretation of the findings and the testing of the hypotheses by the data gathered.

Evaluation

It is difficult to create a structured and objective mark scheme for work which can be very open-ended and which can contain considerable originality of thought. However, the following assessment criteria are used as a guide for marking:

1 Ability to understand those processes which result in the spatial patterns for selected criteria.
Ability to illustrate geographical relationships of measurable phenomena and to account for these relationships, e.g. distance from the centre related to the location of high order services.

7 marks

2 Ability to use mapping and graphical techniques to substantiate arguments put forward.
Ability to rationalise and synthesise information in order to highlight reasons for the acceptance or rejection of hypotheses.

5 marks

3 Ability to recognise the possibility of deviation from the expected and to explain this.
Ability to recognise the limitations of the methods used and to suggest possible alternatives.

3 marks

Pupils at Cleeve have, I believe, substantially benefited from this approach and many average ability pupils succeed in balancing theoretical ideas with the reality of fieldwork findings.

Parts of one pupil's answer – concerned with the hypothesis about main roads and building functions – is shown on p. 37.

The argument is supported by the graph, which shows the order of goods in different zones within the central area. The graph has been constructed from the results recorded on the original base map.

These two graphs show a) How land values decrease as you move from the town centre

b) How land values are affected by main roads

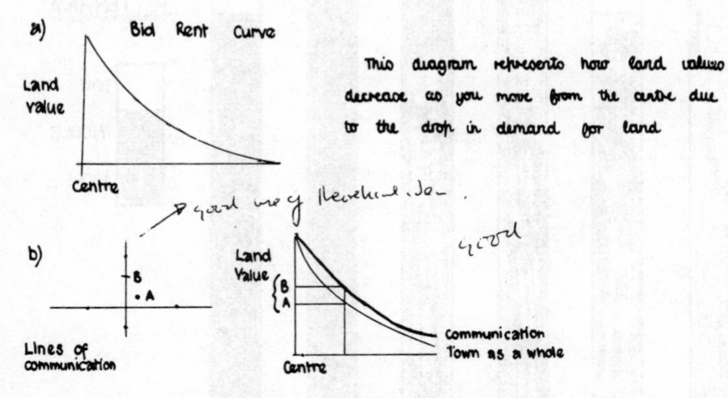

This diagram represents how land values decrease as you move from the centre due to the drop in demand for land

These diagrams show how land value is higher when (as in B) there is line of communication

I have used these graphs to reinforce what does happen to land values as you move from the centre and how land values change with lines of communication. Land values do play a large part in answering this hypothesis.

It can be seen clearly from the map of the functions of buildings in Cheltenham that, with the exception of food services, almost all buildings on the Promenade and the High Street are High Order. As we move from centre and land values decrease so the functions of the buildings change from high to lower order - the graphs prove this is true.

It can also be seen that as we move away from the town centre, especially on the minor roads the buildings are low order, the vast majority are houses anyway. This part of the town is now easily accessible so industries and large shops don't locate there because of this reason. People can afford to live there due to lower land values. I would therefore agree with the hypothesis that main roads affect the functions of buildings and feel that the map presents evidence that it is true.

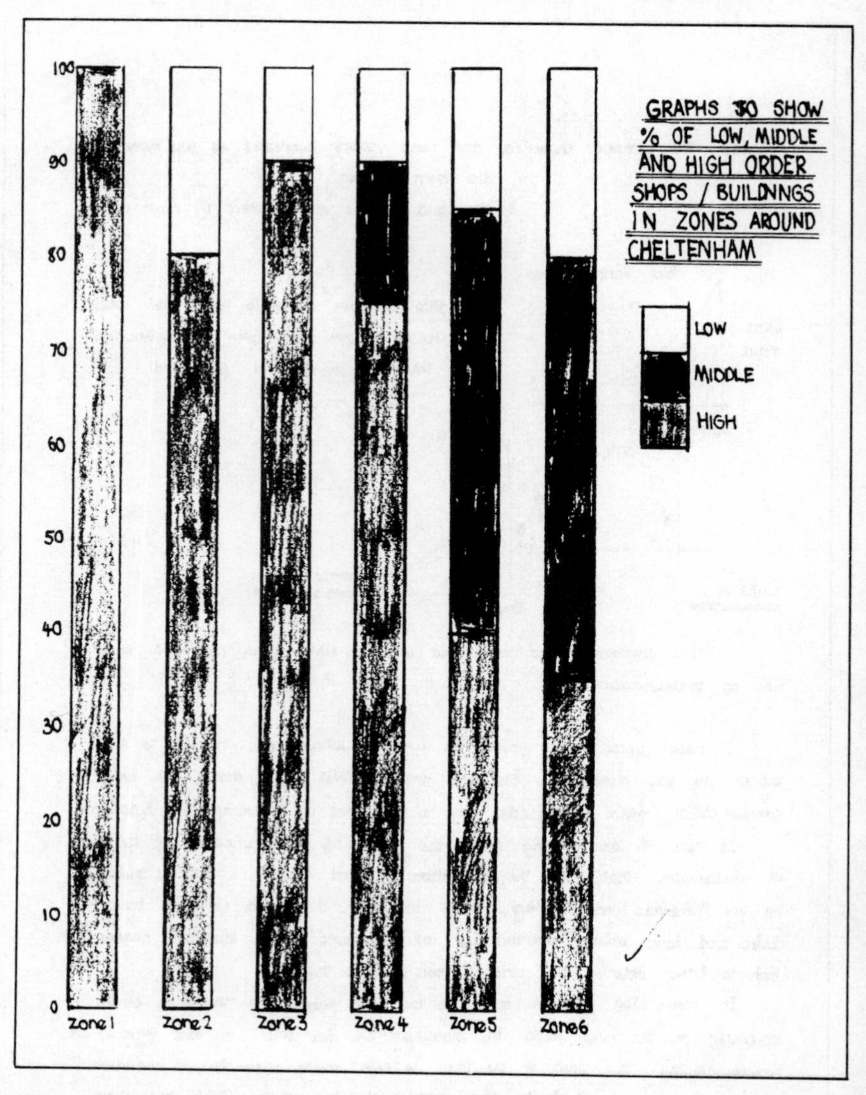

(*From a pupil's original work, using coloured pencils.*)

VII Income distribution – *an operational game*

John Bale

Games and simulations have become a good deal more widespread in school classrooms in the last decade, being used as much to give people an insight into 'the other person's point of view' as to convey factual information about the elements of particular systems.

John Bale's version of an income distribution game begins by looking like an orthodox class quiz – but it reveals its power and interest within a few minutes.

Activities of 'experiential learning' of this kind link closely to the emphases about education which John Huckle speaks of in Part Three.

Introduction

One of the best known examples of how to teach the helplessness of those who feel the system working unfairly against them is the brilliant game STARPOWER devised by Garry Shirts. At the same time participants can 'experience the exercise of total power'. The invariable outcome of STARPOWER is outrage at the unfairness experienced during the playing of the game.

But STARPOWER is rather time-consuming, delicate to handle and ideally requires several dummy runs before operating it. As an alternative, a game with equal potential is presented below.

It is a geographical adaptation of an idea originally suggested by Neubeck in 1979, in the *Handbook for Economics Teachers*, edited by D. J. Whitehead, Heinemann.

Materials and teaching strategies

1 Prepare a ten-question test on any subject, e.g. in a mixed class, *all* the questions might be on football. In a boys' class the questions might be on television – the intention is to aim for a *range* of marks in the class but with some built-in hurdle that skews the marks.

2 Start the test when about 70 per cent of the class are present, so that those who come late miss the first few questions. The teacher might arrange for a colleague to detain some members of the class so that they were certain to miss the first few questions.

3 Students mark their own papers, allowing ten marks for each correct answer. Any student arriving so late that he or she misses the test is given zero. (It might be as well here to assure students that this is only part of the game and that there is nothing to worry about in getting a low score.)

4 Divide the class into three randomly determined groups. The teacher asks each member of each group to call out his or her score and

records these on the blackboard. As this is done, each group should keep a running total of its own group score. This is then added to the blackboard score.

5 At this point the students are told that each *group* represents a country and each *person* with the group represents a region within that country. The total score of the group's test represents the national income and each individual's score represents the regional income.

6 *Task*: To decide how to distribute the income among all of those in the group. The aim of this is to encourage pupils to articulate and think about their values concerning regional economic inequality. It might be useful to suggest some possibilities:
   (i) If the total national income is simply divided by the number of regions each region receives the same amount.
   (ii) The total national income is distributed in accordance with what each region 'earned', i.e. the test score represents value of 'work'.
   (iii) The national income is distributed in such a way that no region receives below a minimum amount.
   (iv) All income should go to one region.

7 The class might be asked to discuss if total regional income could have been increased through co-operation, rather than each student 'working' individually.

Follow-up

The teacher might feel justified in drawing analogues with the real world when discussion is over. It might also be possible to provide a case study of a country which approximates to (i) above, and one which is nearer to (iv).

VIII Slopes, soils and vegetation – *a fieldwork exercise in biogeography*

Graham Corney

In many of the new developments in school geography over the decade, human geography seems to have made the running. Physical geographers, caught unawares by the sudden demise in credibility of the Davisian approach, have retreated to the safe havens of a classification approach to glaciation or arid landscapes. The research in hydrology, on which interest was focused in many university departments, posed technical problems for replication in schools, and so the physical branch languished somewhat – even to the point where some suggested that it might safely be reposited outside the core of the subject.

More recently, concern about physical environmental protection and conservation has revived an interest in integrated man–environment approaches, as exemplified by the chosen strategies of the Schools Council 16–19 Project. And new areas of physical geography have begun to appear in school syllabuses, notably plate tectonics and biogeography.

Graham Corney's fieldwork exercise in biogeography, developed while he was Head of Geography at Lord Williams's School, Thame, is an example of this renewed interest. It represents the investigative hypothesis-testing tradition of the 1970s transferred to the study of interrelationships within the environment, and using student involvement to generate motivation and interest.

Whether scientific fieldwork of this nature will ultimately relate to the humanistic concerns of other units in this book is a matter that the discipline has yet to resolve.

Introduction

The main aim of the exercise was to study some of the interrelationships between slopes, soils and vegetation within a defined environment. In the format that it is described here, the study was carried out on Beacon Hill, part of the Aston Rowant Nature Reserve in the Chilterns, which is in the school's local environment. A hypothesis-testing approach was adopted because of its value in giving structure and focus to the work: thus emphasis was placed on the understanding of key geographical ideas rather than on illustrating discrete facts or relationships, and on the use of techniques as a means of solving a particular problem rather than as ends in themselves. The exercise, therefore, gave students opportunity to test and appreciate some of the interrelationships between slopes - in terms of angle of slope and amplitude of relief; soil quality - in terms of depth, pH value, humus and water content; and vegetation quality - in terms of variety and profusion of species. And it allowed students to employ preparatory reading, map interpretation, fieldwork, laboratory techniques and a statistical test in a structured enquiry.

In addition, partly because the exercise was undertaken with sixth form students, the students themselves were able to be fully involved from the outset of planning. Initial discussion focused on content, so that tentative predictions were made and the two hypotheses established. Further discussions established the objectives and procedures to be adopted. In the planning stage, as throughout the study, the role of the teacher was that of guide or consultant, asking pertinent questions to steer discussion and enquiry towards intended objectives. In fact, this active involvement of students was a fundamental aim of the exercise because it was felt that through this approach students would gain a more effective understanding of the geographical ideas and techniques involved, and of the way in which a geographical enquiry may be planned and carried out.

Materials and teaching strategies

A student task sheet containing instructions and procedure and a data sheet for recording information were compiled. Particular tasks were shared out among the group of students. A brief summary of the main practical tasks is outlined below.

1 The hypotheses decided upon were:
- (i) Soil quality decreases with increasing slope and altitude.
- (ii) Quality of vegetation decreases with increasing slope and altitude.

2 Work in the field was as follows:
- (i) A line transect was set up from the base of the scarp to just over the crest. The transect line was selected from studying maps and from field observations so that it was representative of the profile and alignment of the scarp and of the flora and fauna of the Chilterns. The line was marked by tape.
- (ii) A field sketch was drawn of the whole area under observation, marking the line of transect.
- (iii) The spacing of locations for data collection was decided and each position was marked by a pole. At each location, data was recorded as follows:
- (iv) The angle of slope was measured by clinometer; fore and back readings were taken for accuracy.
- (v) Altitude was determined from sightings on local landmarks which were marked on the OS map.
- (vi) Soil characteristics were studied:
  - (a) depth was recorded, using an auger;
  - (b) one soil sample was collected for subsequent pH analysis;
  - (c) a second soil sample was collected for subsequent analysis of water and humus content;
  - (d) soil temperature was recorded, using a soil thermometer;
  - (e) air temperature was recorded, using a standard thermometer.
- (vii) Vegetation characteristics were studied using a quadrat with grid. The name/species of every plant actually touching each intersection within the quadrat was recorded, identification having been established from keys where possible. Plants difficult to identify were labelled, placed in polythene bags and taken back to school for identification.

3 The follow-up work undertaken included the tasks itemised below:
- (i) An accurate slope profile was drawn, with transect locations marked.

(ii) The field sketch was tidied up and refined.

(iii) A pH meter was used to assess the values of soil samples.

(iv) The water content of samples was established using the following procedure:

(a) samples reduced to standardised weight;
(b) placed in crucible in oven;
(c) left overnight at temperature of 100 °C;
(d) removed, cooled and weighed;
(e) water content established using formula:

$$\% \text{ water in soil} = \frac{C}{A} \times 100$$

where A = weight of soil sample
B = weight of dry soil sample
C = A – B.

(v) The humus content of samples was established using the following procedure:

(a) the same samples were used as for water;
(b) placed in crucible;
(c) heated by bunsen to red heat;
(d) cooled and weighed;
(e) humus content established using formula:

$$\% \text{ humus} = \frac{E}{A} \times 100$$

where A = weight of dry soil sample
D = weight of soil without humus
E = B – D

(vi) Plant identification was completed, and percentage dominance calculated for each location.

(vii) The completed data sheet was studied and the results under each heading were discussed by the group as a whole.

(viii) Indices for the quality of soil and vegetation were selected, and then Spearman's Rank Correlation Coefficient was applied to give statistical accuracy to acceptance or rejection of the initial hypotheses.

(ix) After further discussion, written summaries of results were made, including graphed presentation where appropriate; conclusions on the validity of the hypotheses were stated; and each participant was asked to comment critically on the methods adopted, and to make suggestions for further studies.

4 Equipment used was as follows:

(i) preparation: 1:25 000 maps

(ii) in the field: metre tapes, poles and markers, clinometers, soil auger, polythene bags, labels, soil themometer, standard thermometer, quadrat, ecology keys.

(iii) in the follow-up work: pH meter and distilled water, crucibles, gauges, tripods, bunsens, oven, pocket calculators, critical values for Spearman's Rank Correlation Test.

Both students and teacher felt the exercise considerably helped participants to appreciate the interrelationships specified, to use a variety of geographical techniques in a structured enquiry, and to give considerable scope for student involvement and participation. One factor which helped the exercise achieve its key objectives was that it had developed from previous fieldwork. Earlier versions had been limited in content (for example, a vegetation transect), in techniques (for example, soil characteristics limited to depth and pH value) or in procedure (with little student involvement in the planning stage). The exercise was built, therefore, on earlier versions. As a result, planning and organisation was more effective, and the move towards integrating the three factors in a hypothesis-testing approach was seen to have considerable benefits for student understanding and enjoyment. Another critical factor favouring success was that several of the students involved had worked with the teacher for periods of up to four years. The students, therefore, already had experience of enquiry-based approaches and fieldwork techniques. They were able to play a major part in the planning and execution of the exercise, to evaluate the objectives selected and the techniques employed, and to comment critically on the geographical ideas investigated.

Obviously, however, problems occurred and the predicted results did not all materialise. The vegetation survey was very time consuming in relation to the actual results obtained - although this can often be the case with fieldwork and research. Similarly, though the slope profile comprised a convex upper section and concave lower section, amplitude of relief was only 83 metres and this limited variations in soil and vegetation characteristics. However, while the actual results would probably have been more satisfactory if the exercise were carried out in an area of more active relief, the advantages of being able to work in the local environment were considerable - and, in the event, a statistically significant positive correlation between angle of slope and pH was recorded, and a positive correlation between angle of slope and humus content was only slightly outside the 95 per cent level of significance. Certainly, the co-operation and assistance of the Warden at Aston Rowant was of considerable advantage: access was encouraged by the Nature Conservancy, which is keen to see its reserves used for educational purposes.

And, on reflection, this whole activity produced ideas for further study. A rewarding approach, appropriate to many A-level syllabuses, including that of the Schools Council 16–19 Project, would be to study an area like

Aston Rowant as a case study of an ecosystem. With such an approach, investigations could be made of the roles of climate and energy flows in addition to slopes, soils and vegetation; of the importance of man within the ecosystem; and of the concept of ecosystem management, a key issue for geographical study.

Further reading

BENNETT, D. P. and HUMPHRIS, D. A. (1974) *Introduction to Field Biology* Arnold

HANWELL, J. D. and NEWSOM, M. D. (1973) *Techniques in Physical Geography* Macmillan

MCCULLAGH, P. (1974) *Data Use and Interpretation* OUP

IX Drawing boundaries around a conurbation – *an exercise in 'new' regional geography*

David Lambert

Although the classification of parts of the earth's surface by 'natural' or 'climatic' regions is something of a nostalgic, if not mysterious, activity, there is little doubt that the region has not disappeared for good, as some supposed it might.

Using administrative (e.g. planning board or political boundary) or cultural or communication regions (e.g. as defined by significant telephone-call flows) there is still considerable activity in the study of the earth's surface by area, as opposed to topic or system. But the emphasis seems to turn as much towards the questions of definition as towards what is found within the area.

David Lambert's exercise uses the recent reorganisation of local government boundaries as a springboard towards the consideration of the 'geographical good sense' of such boundaries and towards the very practical problems facing schools and voluntary organisations as a result. It is a regional geography exercise rooted in the practicality of a real and local problem, and one which can readily be replicated in local study in almost any part of the United Kingdom.

In facing the problem, it tackles some fundamental geographical concepts (catchments, territory, boundaries) and uses map study and skills in working towards an understanding of the rather more complex problem of finding appropriate spatial delineations for *groups* of functions.

The lesson was taught originally to a lower sixth form, but is readily usable with younger students.

Introduction

The matter of defining regions and drawing boundaries around regions has not exactly been central to school geography recently. It is, though, a problem that faces us all; how to organise the space in which we live. This can be taken from the trivial household level to the problem governments have in organising their territory for efficient administration.

The bases for organising Britain into administrative regions are the *counties* (County Councils) and *districts* (District Councils). The former have responsibility for services such as Education and the latter Housing (see Fig. 7). The boundaries of these administrative regions have not been

	Planning	Education	Environmental Health	Housing	Recreation	Traffic
County	✓	✓			✓	✓
Districts	✓		✓	✓	✓	
Metropolitan county	✓				✓	✓
Metropolitan districts	✓	✓	✓	✓	✓	
GLC	✓			✓	✓	✓
London boroughs	✓	✓	✓	✓	✓	

*Note: this is a generalised list, and individual responsibilities under any heading may show slight differences.

Fig. 7 *Responsibilities of different local government units* (*England and Wales only*)*

with us long, however. Before local government reorganisation in 1974, we still had counties (with Urban and Rural Districts within them) but their boundaries were very different, having been established in 1888 by the Local Government Act. In fact, the large county boundaries had been with us for a very long time indeed, evolving as they did well before the Industrial Revolution.

The aim of this exercise is to examine an important problem in geography, namely drawing boundaries around regions on a map, using local government reorganisation as a practical example. Although the government introduced the new boundaries several years ago now, it still remains a contentious issue; many people complain that the new counties are too large, too remote and that local government is no longer 'local'. Others, particularly planners, have criticised the new boundaries from completely the opposite point of view. Perhaps one of the results of this exercise will be to show that there are no 'perfect' answers; a boundary has to be drawn, it has to go somewhere and it is bound to upset someone.

When teaching this unit it has been found that an introductory discussion on local government must be undertaken. It seems that, even though local government pervades all our lives, it is a very 'taken for granted' issue and students even of sixth-form level are not at all sure of its function, let alone the specifically geographical problem of defining local government areas. Under some circumstances this can become very wide ranging, and some difficult but essential questions are thrown up such as:

> What are regions? Are they merely the result of geographers *classifying* space? Or do they sometimes have a *function*? What are the functions of counties and the smaller, more numerous, districts?

This discussion can be controlled and steered toward the more specific concern of why reorganisation was necessary in 1974. This can quite readily take in many geographical concepts and provides a forum in which terms and ideas such as mobility, spheres of influence or urban–rural differentiation can be used in a practical sense. The old local government areas, for example, were increasingly unrealistic. County boroughs were the local administrative units of the main urban areas and the urban and rural districts were the subdivisions of the old counties; but their boundaries had been drawn when the spheres of influence of towns and cities were restricted (by today's standards) owing to the low level of mobility. Urban areas were also relatively compact and commuting more than walking distance still unusual. These nineteenth century boundaries, therefore, were failing really to organise space realistically in the 1960s because an increasing standard of living and universal personal mobility had precipitated many changes in the way people behaved. Not least it led to the physical spreading outward of the towns and cities by suburbanisation and also to a huge increase in the size of their spheres of influence.

Similarly, before the exercise is undertaken, the present day boundaries ought to be examined in general. For the first time in centuries the actual county boundaries experienced wholesale change; some counties (e.g. Rutland) disappeared and some new counties appeared (Cleveland). Counties are subdivided into districts. An important new departure was the creation of the metropolitan counties (and metropolitan districts) based upon the major cities, and here is a clear opportunity to discuss the growth of the conurbations and perhaps the complex patterns of movement and interaction between the parts of the conurbations. Indeed, this is the time to introduce the contentious point about the boundaries of the metropolitan counties; that 'these metropolitan authorities had boundaries which most commentators have regarded as too tightly drawn, thus infringing on the city – hinterland principle' (House, p. 25[1]). Clearly the major problem was to define the real extent of these conurbations for the purposes of efficient administration and it is certainly within the experience of many sixth formers to appreciate that commuting hinterlands extend far beyond a line drawn around the continuous built-up area of a large city or conurbation. But how far can a conurbation reasonably be said to extend into the surrounding countryside?

If your school is situated in such 'surrounding countryside' then, emotionally, pupils will surely justify their non-metropolitan status. In fact several areas, located a little too close for comfort to the conurbation, fought proposals to include them in the metropolitan county in question. Poynton, for one, was successful in winning a reprieve and remains in Cheshire mainly on the grounds that its regional centre was Macclesfield (see Figs. 8 and 9). The exercise here does not deal with Poynton *per se*, but aims to show that this part of the compromise boundary between Greater Manchester, Cheshire and Derbyshire is in some ways deficient; it is insensitive to local geographical conditions (both of a human and a physical nature) and perhaps too sensitive to local opinion – a discussion point on its own!

Materials and teaching strategies

Pupils will require copies of the maps (Figs. 8 and 9), plus the explanation and questions sheet. The pupils will use the maps to investigate how local government reorganisation has affected the organisation of two aspects of local life, namely secondary education and the Girl Guides movement. Some discussion points are noted afterwards and it would not be stretching a point too far to broaden the discussion to include the whole problem of urban–rural differentiation.

Some discussion points

The catchment areas drawn on the maps can of course be compared with your school catchment area and individual contrasts and similarities can be identified. Points worth emphasising are how the pre-organisation county boundaries themselves, and the road network (itself related to some degree to the relief of the area), influenced the location of secondary schools in Marple. Marple became the natural focus in so far as secondary schooling was concerned, and the town also took on the same role for Guiding, being the 'centre' of the Chadkirk Division.

However, after reorganisation, Marple became part of the Stockport Metropolitan District (a district of the Metropolitan County of Greater Manchester), leaving Disley to remain in Cheshire (this suited Disley

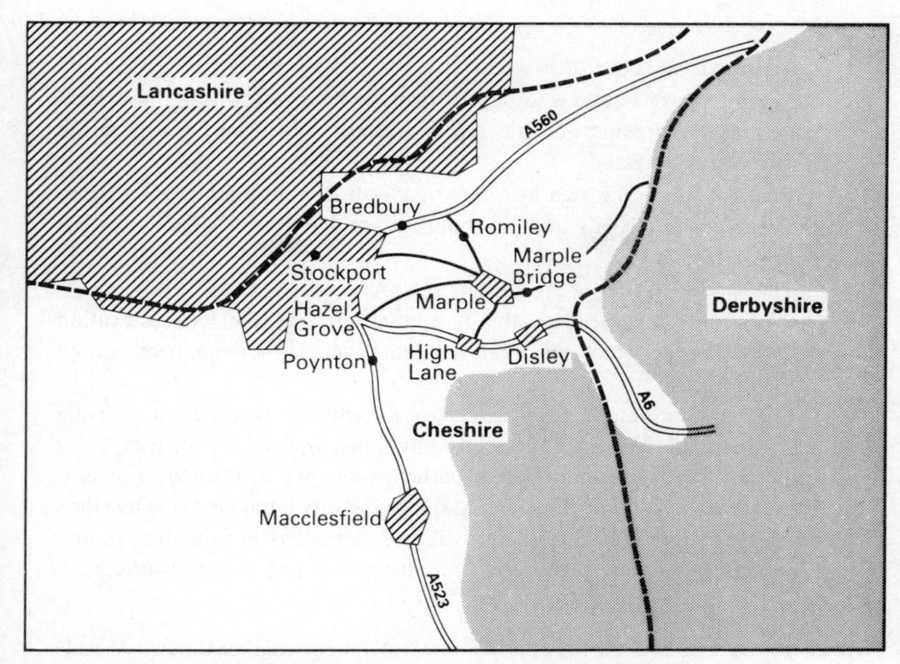

Fig. 8 *The south-east Manchester area before reorganisation*

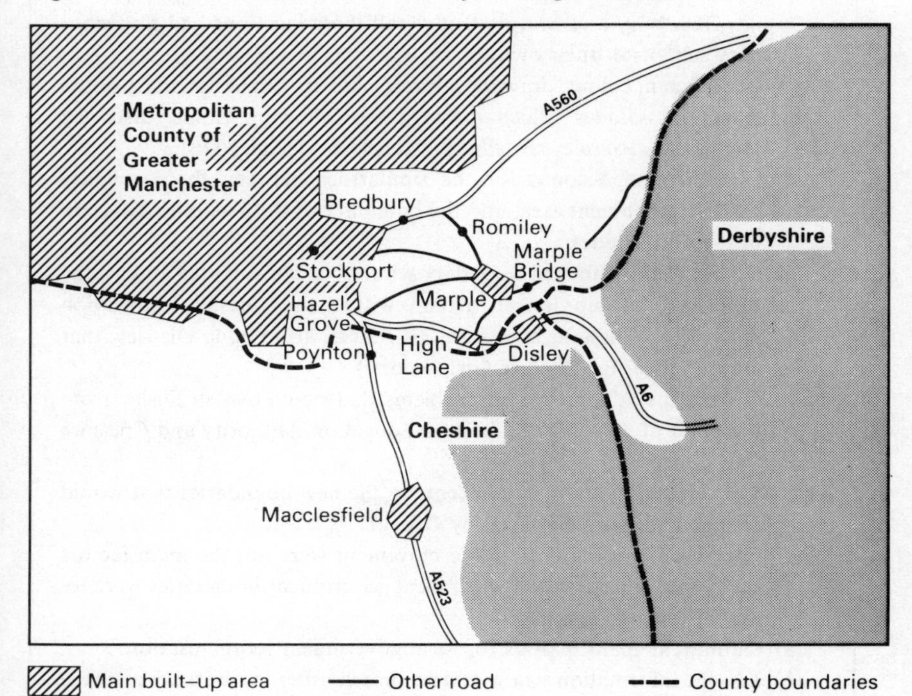

Fig. 9 *The south-east Manchester area after reorganisation*

INFORMATION AND QUESTION SHEET

The aim of this exercise is to use the maps to help you consider how local government reorganisation has affected aspects of life in an area in the north-west of England.

Figure 8 shows the area before reorganisation – the county boundaries, the relief, major roads and the built-up area. Figure 9 shows how the boundaries changed.

The aspects of life we are considering are:

(a) Secondary education; this is administered by the local education authorities which are organised under the counties and metropolitan districts.

(b) The Girl Guides; administered on a county basis, the counties being subdivided by the Guides into 'divisions' and smaller 'districts'.

These two activities both depend, either wholly or partly, on local government money. It is clear, therefore, that if the county boundaries change then many schools, or Girl Guide units, will find themselves in a different county – looking to a different authority for educational policy and resources.

Questions

1 (a) On Fig. 8, draw the approximate 'catchment area' of the two Marple secondary schools. It included Disley, High Lane, Hazel Grove, Romiley, Bredbury and Marple Bridge (all these locations had no secondary schools of their own).

(b) In a different colour, draw the approximate divisional boundary of the 'Chadkirk Guides' which included the following Guide 'districts': High Lane, Romiley, Bredbury, Marple Bridge and Disley.

(c) Describe and account for the similarities between the secondary schools' catchment area, and the boundary of the Chadkirk Division of the Girl Guides.

2 (a) On Fig. 9, draw the old secondary schools' catchment area and list the locations that no longer remain within the Cheshire Education Authority. (These locations are, of course, also Guide Districts that are no longer in Cheshire Girl Guides.)

(b) Outline the administration problems that now exists in Disley from the point of view of the Cheshire Education Authority and Cheshire Girl Guides.

3 Can you suggest any adjustments to the new boundaries that would alleviate the problem faced by Disley?

4 Education and the Girl Guide movement were not the main factors taken into consideration when local government boundaries were redrawn.

(a) Outline the main reasons for local government reorganisation.

(b) List the information you would require in order to assess whether the boundaries shown in Fig. 9 are realistic. In other words, does the boundary of the county of Greater Manchester reflect the true extent of the Manchester conurbation? What information would you need to answer this question?

residents at the time). Should Disley have been included in the county of Manchester also? This is a question taken seriously by some residents because, as the exercise will have shown, Disley became very isolated. It had looked toward Marple but now Macclesfield became its administrative centre – even though historical and physical factors really militated against this. More seriously, there was no easily accessible secondary school for Disley children inside Cheshire; nor was there a sufficient catchment area around Disley to warrant building a new school in Disley. The Cheshire Girl Guides also had a problem, Disley really being out on a limb. Thus to include Disley in the Metropolitan County might have been more satisfactory.

In the event though, certain *ad hoc* measures were taken to help rationalise the situation. Stockport MD, having acquired the Marple schools, had too many schools and so agreed to take Cheshire CC children from Disley, who were bussed in just as they were before. The Guides were more radical and decided to organise their divisions *across* the new county boundaries, this making sense from a local administrative point of view. For example, Disley now forms a district with one of Stockport's divisions, running along the A6 trunk road and including High Lane and Hazel Grove. It is, of course, only a compromise, for although administratively Disley belongs to Stockport Guides, from the point of view of local government money (for camp training or equipment) Disley is still part of Cheshire!

Reference

1 HOUSE, J. W. (ed.) (1977) *United Kingdom Space*. Weidenfeld and Nicolson

---

X Teaching welfare issues in urban geography – *a work unit on externalities*

John Bale

Unit I (Neville Grenyer's Industrial Revolution location exercise) has already touched on the theme of 'externalities' – the spillover effects of a particular decision on those who may be unaware or powerless to intervene in the situation.

John Bale here offers a major unit of work which considers the whole topic within the context of the 'welfare approach' to geography. Books by D. M. Smith[1], and by Coates et al.[2] represent accessible basic texts for teachers in this area and Bale is concerned that the introduction of welfare themes and approaches should bring school geography closer in spirit to much of the work currently being undertaken at tertiary level:

'Students need to be aware of the inequalities in society and they need to arrive at informed value positions with regard to these inequalities. In doing so they ought to undertake interesting, relevant and motivational work.'

Bale's unit begins with a simple example of an unwanted 'externality' and moves on to develop a classification of the types of effect that can be created. Students progress to an exercise in which they consider how they would react to certain effects in their own neighbourhood. The third and fourth parts of the unit include a mapping exercise in which traditional skills are required, and a piece of urban fieldwork.

As presented here, the work is geared for sixth forms, but the author has, in a modified form, used it with younger groups. The full unit might take as long as twelve lessons to complete, but the exact time would depend on student and teacher enthusiasm and interest.

Introduction

It could be argued that the concept of externalities is the most important, though least taught, in the whole of human geography. Externalities occur when the behaviour of one individual, firm or facility affects the welfare, positively or negatively, or others. Mishan[3] notes that 'the operations of firms, or the doings of ordinary people, frequently have significant effects on others of which no account need be taken by the firms, or the individuals responsible for them'. Clearly, the *location* of an individual with respect to the source of the externality is absolutely vital in assessing the individual's level of welfare. Hence, the 'externality field' (or the sphere of influence of the source of the spillovers) is a key concept in welfare geography.

1 *A classification of externalities*

The work unit might commence with the teacher introducing the idea of 'spillovers' from day to day examples which everybody experiences. The following letter in The *Guardian* is a useful example (Fig. 10).

Having examined the variety of spillover effects which exist, the teacher might suggest that they can be classified in the following way, based largely on the work of Harrop[4]

(a) *Positive or negative*. This is the most basic and most important dichotomy. Externalities are negative when 'people consume more of them than they would freely choose' (D. M. Smith[1]). A simple example of a negative externality is the case of my next-door neigh-

Blue note

Sir, — Last week you reviewed events at the Alexandra Palace Open Air Jazz Festival. While its artistic success was undeniable, with all-star casts giving excellent entertainment, for local inhabitants it was an environmental disaster.

Even from half a mile distant, noise levels were intolerable, with a jumbled cacophony of distortion insufferably disturbing a normally quiet, residential area.

Surely such open-air music festivals must in future be restricted to non-residential areas. Alexandra Palace remains a vexed question in the minds of the planners, but its natural amphitheatre is simply too near housing to make this an environmentally acceptable answer. — Yours faithfully,
Brian Wright
34 Priory Road,
London N8.

Fig. 10

bour who allows his back garden to degenerate into a weed-infested plot. My quality of life is reduced because of the visual intrusion of his untended plot. Other obvious examples of negative externalities are airport noise, pollution from a factory or smells from a sewerage plant. Pollution, in fact, is a classic example of an externality 'because in most cases the generator of pollution does not have to pay the costs it imposes on society, partly because the media which are polluted – the air, water supply, the sea and the beach, open spaces – are public goods, that is non-marketable goods enjoyed by all' (Richardson[5]). The costs are, in fact, borne by the unwilling consumer. The facilities which create negative spillovers can also generate those of the positive variety. The airport and the factory which are polluters also provide jobs for local residents. Towns which create congestion also provide shops for a large number of people both inside and outside the town. A useful distinction to note at this point is that negative spillovers affect those who are *proximate* to their sources, whereas the positive variety benefit those who are *accessible* to the source. Negative externalities are obviously more serious than the positive kind because people have no choice in consuming them.

(b) *Urban or rural.* In the countryside 'space has the ability to isolate or internalise potential externalities' (Harrop[4]). The smell of a pig farm in the country will affect a tiny number of people compared with the number it would upset if located in the middle of a city. For this reason externalities are more serious in urban than in rural areas.

(c) *Perceived and unperceived.* Some externalities are obvious to us; we can actually see air pollution, we can smell odours from a molasses factory. Other externalities may be equally (or more) serious, but may be unperceived by people who are exposed to them. The gradual permeation of the atmosphere by low levels of nuclear fallout around a nuclear reactor is a case in point (Bunge and Bordessa[6]). However, other negative externalities may be so minimal as to be harmless. The question of deciding when a spillover is, or is not, negative in nature might in many cases be determined by whether it is perceived as such (see below).

(d) *Inter-use and intra-use.* In the intra-use category, externalities affect areas of the same land use – factory noise in an industrial estate, for example. In the inter-use category externalities are between different types of land uses, a factory affecting a housing estate, for instance. Obviously, this latter case is the more serious.

(e) *Scale differences.* Some spillovers extend over very large areas, others only very small distances. The sphere of influence of a town (a positive externality) may extend over many kilometres; the visual intrusion of my neighbour's garden over only a few metres.

The above typology may not be complete, nor are the categories mutually exclusive. The classification does, however, make a useful starting point for a work unit on externalities. The teacher, having introduced the concept by class discussion, may invite students to think of their own examples of each type and consider, for each case, which is the most serious. This exercise could be undertaken on the accompanying worksheet (Fig. 11).

2 *Identifying neighbourhood attributes and facility preferences*

This part of the work unit is designed to demonstrate that the pattern of facilities which generate positive and negative externalities, and thus create indirect costs to (unwilling) consumers, 'varies quite substantially through the urban system, so that some groups go fairly cost free while others suffer very considerable burdens' (Harvey[7]). The problem of actually putting a money value on the costs incurred by those who consume negative spillovers is very difficult. One thing is clear, however. 'The cost to the individual in each case will be a function of his [sic] location with respect to the generating source' (Harvey[7]).

Type	*Example*	*Which is the most important and why?*
Positive		
Negative		
Urban		
Rural		
Perceived		
Unperceived		
Intra-use		
Inter-use		
Large spatial impact		
Small spatial impact		

Fig. 11 *Worksheet for exercise on classification of externalities*

The attributes of a particular neighbourhood may not be those that residents would freely choose. As urban change occurs long-time residents of particular areas may suffer in their quality of life as a result of development and urban change. A motorway running through a neighbourhood can generate major environmental downgrading as well as psychological stress for residents. A public library, on the other hand, might be welcomed by local residents since it probably only generates positive externalities.

The distribution of different facilities in an urban area is such that certain social groups consume more negative externalities than others. In that such spillovers are unpriced and impose costs on the groups in question, various 'fringe benefits' and 'fringe disbenefits' of the urban system are distributed unequally across the urban population (Harvey[7]). In order to teach these ideas the following steps might be undertaken with a class.

(a) Students are presented with a table of facilities (Fig. 12) which they are invited to score on a 'pride–stigma' continuum (Wolpert et al.[8]). Students look at the list of facilities and, assuming that they (the students) are permanent residents of their neighbourhood and that any one of the facilities may be introduced into their neighbourhood or into a neighbouring area, they score each facility on a scale from 1 (highly desirable) to 7 (highly undesirable), with 4 indicating indifference.

(b) Beside each facility which the students have scored 5, 6 or 7 they are asked to indicate how they would respond if the facility in question were to be located in their particular area. This is done by using one of six letters as follows:

A – you would move elsewhere;
B – you would actively protest, or organise a resistance group;
C – you would join an already organised resistance group;
D – you would complain but not act;
E – you would do nothing;
F – other (specify).

(c) Having undertaken this exercise individually, a group score can be obtained for each facility and each location, i.e. all scores for each cell in the matrix are summed. Lowest total scores refer to those facilities which people would actually welcome in their neighbourhoods because of the positive externalities which they provide. High scores will refer to those unfavourable facilities which create negative spillovers. The fact that some facilities might be unwelcome within, say, 40 metres of a residence but would be welcome within the neighbourhood as a whole demonstrates how sensitive individuals are to the *distance* from the facility in question.

(d) The next stage in the study of externalities is to establish *where* in the urban area the favoured and unfavoured facilities are, in fact, located. Do those facilities which are generally favoured cluster in particular parts of the urban area? Are the unfavoured facilities concentrated in the central city? Depending on the time available any number of high and low scoring facilities can be located on a map of the nearest large urban area, either directly from a land use map or from a telephone directory and street plan.

The spatial patterns depicted from a simple mapping exercise such as that described above will reveal the location of 'goods' and 'bads' in the urban area. Whether the system leading to the pattern depicted is right or wrong is, of course, a value judgement. Nevertheless, students should be invited to discuss the results of the exercise and possibly suggest alternative urban plans which would eliminate some of the injustices apparent from the mapping exercise. In doing so they might well arrive at a planner's solution to the problem of externalities, that of zoning different urban land uses so that inter-use spillovers are minimised.

3 *Mapping externality fields*

In the classification of externalities presented earlier, it was noted that some may extend over many kilometres, with others spilling over only a few metres. The sphere of influence of a utility may be presented theoretically as in Fig. 13, the impact of the externality being seen to attenuate with distance from its origin. If the 'externality gradient' is rotated through 360° a sphere of influence is delimited. Theoretically, the exter-

Facility	*Within 40 metres of your house*	*In your neighbourhood*	*Within a neighbouring area*
1 Hospital			
2 Office block			
3 Police station			
4 Museum			
5 Large factory			
6 Cemetery			
7 Filling station			
8 Supermarket			
9 Nightclub			
10 Football stadium			
11 Church			
12 Cafe			
13 Park			
14 Launderette			
15 Theatre or cinema			
16 Bank			
17 Oil refinery			
18 Amusement park			
19 Urban motorway			
20 Sewage plant			
21 Comprehensive school			
22 Antique dealer			
23 Chinese take-away			
24 Fire station			
25 Salvation Army hostel			

Fig. 12 *Facilities in the pride–stigma continuum*

nality field will vary in intensity and, of courser, in shape. Rather surprisingly, Harvey[7] avers that 'we know very little about the shape and form of these externality fields in an urban environment'. In fact, externality fields of many facilities have been mapped by geographers for many years. The spheres of influence of towns, of factories, of industrial location factors and of schools are all examples of positive externality fields, the utilities which generate them providing benefits for those with access to them.

Negative externality fields too have been investigated in considerable detail. Starkie[9] illustrates the spatial impact of various forms of pollution and a useful review of the kinds of mapping pollutants, the nature of the damage they make and the ways in which they are measured, has been provided by Lakshmanan and Chatterjee.[10] Such objectively delimited externality fields as that produced by airport noise or lead

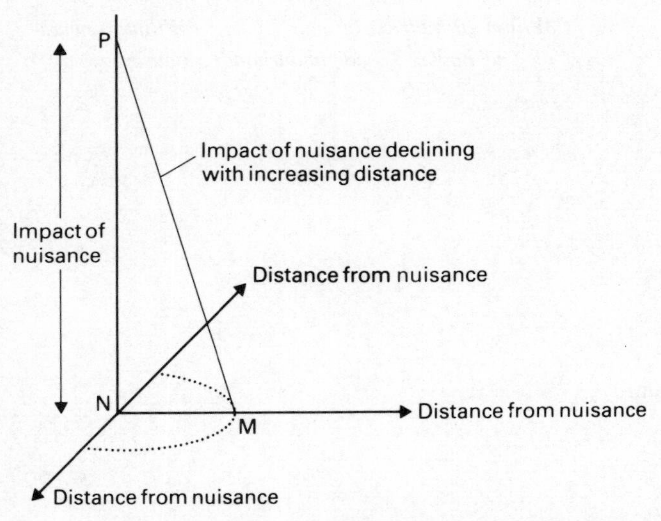

Fig. 13 *The theoretical delimitation of the sphere of influence of a nuisance. N–M indicates the radius of the sphere of influence*

concentration in the atmosphere may be used to illustrate to students the idea of the spatial extent of such negative spillovers. They are, however, inappropriate for fieldwork for two reasons; one financial, and other methodological. The scientific instrumentation necessary for recording noise or dirt levels would be beyond the financial resources of all but the wealthiest schools. Furthermore, the delimitation of a negative externality field of, say, an airport, would be very time consuming. A second reason for the inappropriateness of 'scientific' approaches is that beyond a critical level – the 'net negative externality field', perhaps (Bale[11]) – the impact of the nuisance while measurable scientifically, is not perceived as a nuisance. 'A certain level of pollution or environmental nuisance can be tolerated without harmful effects' (Richardson[5]), even though the pollution may be detected by scientific measurement. Hence, a perceived negative externality field might be the most appropriate to map in the field, for it is what people *think* is a nuisance that is usually more important than the presence of very low level of scientifically measured pollution.

4 *Delimiting negative externality fields*

The aims of the fieldwork session described in this section are fourfold. First, to delimit the spatial extent of the externality fields of several nuisances in the local urban area. Second, to establish the elements which make up the nuisance, ranking these elements in order of perceived importance. Third, to establish whether residents living within the 'nuisance field' obtain compensation or not. Fourth, to critically examine the method of delimitation and note its shortcomings and possible ways of improving it.

The delimitation of the 'nuisance fields' might proceed as follows:

(a) In the classroom, students identify a number of nuisances in the urban area which might be expected to generate negative spillovers. The present author's students have investigated such nuisances as factory areas, waste tips, football grounds and public houses. Other obvious utilities which might be investigated include sewage plants, airports and stretches of urban motorway. The number of nuisances investigated will obviously depend on time and numbers of students.

(b) Having identified the sources of negative spillovers, sample points are located within about 2 kilometres of the sources. This would be ideally undertaken on a 1 : 10 000 Ordnance Survey map or a local street plan. Each sample point should fall on a residence. Fifty sample points around each source would probably represent the minimum and if students worked in pairs this would be a realistic number to be interviewed during an afternoon's fieldwork.

(c) In the field, residents at each sample point are interviewed briefly. This interview method of investigating externalities in the field contrasts with the approach described by Hammond,[12] but is nevertheless thought worthwhile in the present context. Politeness and courtesy from students are obvious prerequisites, however. The following questions are put to the interviewees:

    (i) Living here, do you find .... (name of facility)
    1 A serious nuisance;
    2 A nuisance;
    3 No nuisance at all.
    (students tick appropriate response)

    (ii) If 1 or 2, in what way is it a nuisance?
    (students list nature of nuisance)

    (iii) Do you receive any compensation from the source of the nuisance?

    (iv) Do you receive a reduction in your rates as a result of your proximity to the nuisance?

    The wording of question 1 might need to be altered according to the nature of the nuisance. For example, a visual intrusion might be investigated by asking whether, living at the sample point, the respondent finds it 'a serious eyesore', 'an eyesore', or 'no eyesore at all'.

(d) Following each interview students record a score of 2, 1 or 0 on a base map at the appropriate sample point, depending on whether the respondent answered 1, 2 or 3 respectively to question (i).

(e) On returning to the classroom students map their results to question 1. By interpolating two 'nuisance contours', one delimiting the 'serious nuisance' responses from the 'nuisance' responses, an indication of the perceived intensity of the nuisance field can be obtained. An example of such a map is shown in Fig. 14 which depicts the nuisance field of a factory area at the southern entrance to the Blackwall Tunnel in south-east London.

(f) Students now note the geographic extent of the nuisance field. In Fig. 14, for example, it extends up to 800 metres into the residential area south of the factories.

(g) The results of the other questions can be analysed and tabulated in an attempt to discover the most frequently cited elements of each nuisance and the degree to which compensation is provided.

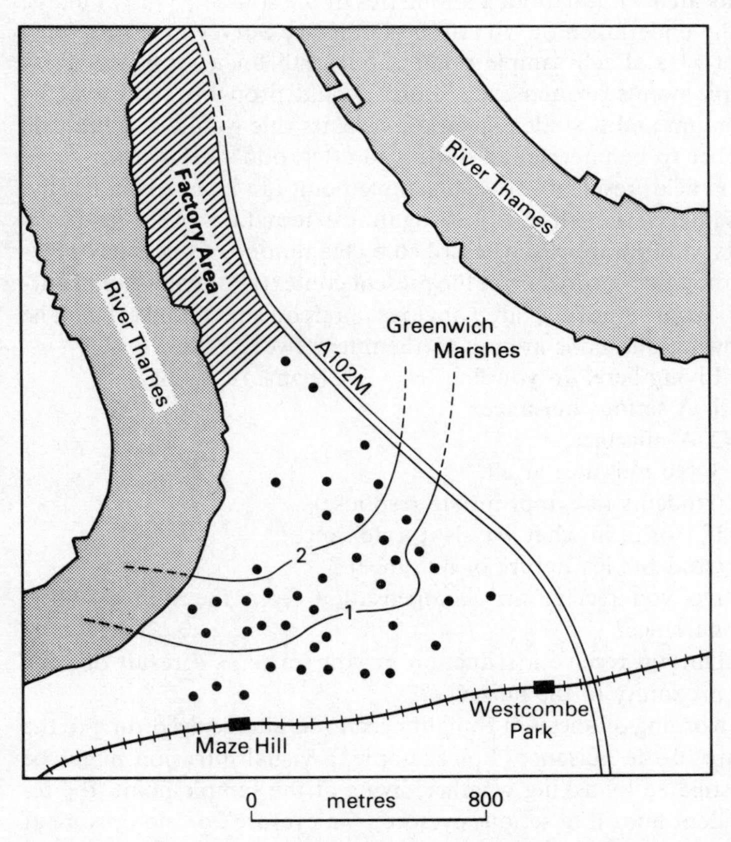

Fig. 14 *The sphere of influence of a nuisance. The spatial extent of perceived nuisances generated by a group of factories in south-east London*

While this method of analysis might appear somewhat crude (what might be a nuisance to some might be a benefit to others), it does possess the advantage of being easily operational. Providing the shortcomings of the method are fully recognised by students, the approach would seem justified.

The component of the work unit will illustrate very clearly the fact that proximity to certain utilities produces negative spillovers for urban residents. At a very general level the exercise has been able to delimit and map the intensity of negative externality fields around a number of urban facilities (for another example, using this method, see Bale[13]).

5 *Solutions to the problem*

How might the problem of negative externalities be minimised? Is it fair that people living near nuisances should bear the costs of the externalities? These kinds of questions might be explored by teacher and students following the mapping exercise described above. Frequently suggested solutions may be briefly mentioned. First, generators of externalities should pay the social costs incurred by those external to the source. This assumes that costs can be objectively measured. In some cases this is obviously possible but in others it is rather more difficult. How can the 'cost' of the mental stress and strain of living next to a busy road be measured, for example?

A geographical solution, implicit in the zoning of urban land uses since the Town and Country Planning Act of 1947, is sometimes described as the 'separate facilities solution' (Mishan[3]) where nuisance sources are isolated or segregated from those they adversely affect.

An alternative solution is one which theorises as follows:

> If an environmental nuisance such as noise is newly introduced into an area then some of the more sensitive households ... will leave that area. With a distribution of sensitive people which is random in the initial case, more people will leave the areas of higher nuisance than areas less affected. Incomers to the area with their assumed perfect perception will correctly assess the effect of the noise environment on housing, utility and thus expect to be compensated by a lower purchase price. (Starkie and Johnson[14].)

Whether this actually happens in practice is highly debatable.

6 *Linked activities*

It is quite possible to take this work unit further, if time was available and if students and teacher felt inclined. A value analysis after the style discussed by Kracht and Boehm[15] is an obvious follow-up to the work already described. In such work students might explore their attitudes and values with regard to pollution and other spillovers. Another activity which would seem desirable in a longer work unit would be a role play in which students identify with different interest groups in a conflict situation involving spillover effects from a particular utility. A good example might be the case of the proposal to drown a river valley to provide water for a distant city. Here the negative externality of the demand for water has effects on local residents whose homes might be lost. A scenario for this example is provided by Sully[16]. Such role plays should teach empathy, as well as indicating how a given utility, or a less tangible example such as 'economic growth' can create both positive and negative spillovers at the same time.

References

1 SMITH, D. M. (1977) *Human Geography: a Welfare Approach.* Arnold

2 COATES, B., JOHNSON, R. J. and KNOX, P. (1977) *Geography and Inequality.* OUP

3 MISHAN, E. J. (1969) *The Costs of Economic Growth.* Pelican

4 HARROP, K. J. (1974) 'Nuisances and their externality fields', *Seminar Paper*, 23, University of Newcastle upon Tyne, Department of Geography

5 RICHARDSON, H. (1971) 'Economics and the environment', *National Westminster Bank Quarterly Review*, May

6 BUNGE, W. W. and BORDESSA, R. (1975) *The Canadian Alternative*, Geographical Monograph, 2, York University, Department of Geography

7 HARVEY, D. (1973) *Social Justice and the City.* Arnold

8 WOLPERT, J., MUMPHERY, A. and SELEY, J. (1972) *Metropolitan Neighbourhoods, Participation and Conflict over Change*, Commission on College Geography, Resource Paper, 16, Association of American Geographers.

9 STARKIE D. N. M. (1976) 'The spatial dimensions of pollution policy', in COPPOCK, T. J. and SEWELL, W. R. D. (eds) *Spatial Dimensions of Public Policy.* Pergamon

10 LAKSHMANAN, T. R. and CHATTERJEE, L. (1977) *Urbanisation and Environmental Quality*, Commission on College Geography, Resource Paper 77-1, Association of American Geographers.

11 BALE, J. (1978) 'Externality gradients', *Area*, **10**(5), 334–36

12 HAMMOND, R. (1978) 'Externalities: the benefits and penalties of where you live', *Teaching Geography* **4**(2), 68–70

13 BALE, J. (1980) 'Football clubs as neighbours', *Town and Country Planning*, **49**(3), 93–4

14 STARKIE, D. N. M. and JOHNSON, D. M. (1975) *The Economic Value of Peace and Quiet.* Saxon House

15 KRACHT, J. B. and BOEHM, R. G. (1975) 'Feelings about the community: using value clarification in and out of the classroom', *Journal of Geography* **74**(4), 198–206

16 SULLY, E. (1975) 'The drowning of the Senni Valley', *Bulletin of Environmental Education* **49,** 15–17

Further reading

HICKMAN, G., REYNOLDS, J. and TOLLEY, H. (1973) *A New Professionalism for a Changing Geography*. Schools Council
KIRBY, A. and LAMBERT, D. (1978) 'Geography at school and University: is the gap between them growing?' *Papers in Education in Geography*, 2, University of Reading, Department of Geography.
SMITH, D. M. (1977) *Human Geography: A Welfare Approach.* Arnold
STODDART, D. (1967) 'The growth and structure of geography', *Transactions, Institute of British Geographers* **41,** 1–20

XI On teaching about values in the classroom – *a lesson and a retrospect*

Liz Ambrose

The question of 'values education' has become a talking-point in many geographical conferences within the last few years. Liz Ambrose explains here how the question of pupil value judgements about foreign aid became a focus in an orthodox preparing-for-A-level lesson with a sixth form group.

Her own experience and reflection on this lesson led her to subsequent work in searching out references, on ways of teaching about values. She goes on to describe a possible reformulation of the lesson in the terms of a 'values clarification' approach.

This is one of a number of approaches to values education in the classroom which John Huckle discusses in his comprehensive chapter on this topic in Part Three.

Introduction

John Eyles has written: 'What is needed is the adoption of values that represent our aspirations for social change; in other words, a fundamental reform of the social and spatial systems in favour of the disadvantaged.' P. T. Newby, discussing this, comments that 'it is doubtful whether [these] objectives would be acceptable or even desirable at school level' (*Geography*, **65,** (1), 13–18).

It is my belief that at A level at least, when we are working with young adults, Eyles's approach is highly desirable. If we do not identify with the objectives which he states, we are in danger of presenting a picture of the current world situation in its status quo, *with the tacit approval which that implies*. I believe that as teachers we should make it clear that we are concerned about the 'fundamental reforms' which Eyles mentions and that we should guide our students to the data necessary for them to make up their own minds. They can then accept or reject our concern.

A lesson experienced with a second-year A-level geography group, preparing for the Paper III of the London Board in early 1980, gives rise to these notes. On the surface, the fifty minute session was not especially remarkable, since it consisted of a discussion which followed a rather roundabout route.

However, it proved very revealing to me at the time. It has made me consider, in retrospect, the need to confront pupil value judgements and incorporate them into learning programmes.

Materials and learning strategies

Aim of lesson

To investigate the role of foreign aid, in its different forms, to countries of the developing world.

Objectives

To examine the different types and levels of aid given by particular donors and to discuss the long-term and short-term usefulness of such aid to the recipient countries.

Materials

Data on aid totals, plus increases and decreases at a percentage of budget totals, of Northern donor countries over a period of years. Editorial and articles on the functions of aid from *New Internationalist*, December 1979.

Discussion quickly revealed three sets of initial opinions held by pupils.

1 Aid is pointless, since all the recipients do is to take the opportunity to increase their populations as a result. We are thus throwing good money after bad into a bottomless pit, and some sort of Malthusian check would be a very good idea right now.
2 'Charity begins at home'. In the current financial situation of this country and many other countries of the Northern hemisphere budget priorities should aim at improving employment and other chances at home. Aid programmes should be dropped altogether.
3 The egalitarian view. Everyone has a right to different life-chances and so the privileged countries have a duty to share their wealth with the under-privileged. Aid must be given in large quantities until equal wealth exists amongst countries.

We then examined the available data on different types of aid, and considered a handful of case studies which covered voluntary aid, government aid, bi-lateral aid, aid with strings attached, and etc. We came to the end of the session with some slightly revised views and certainly a more informed basis of judgement.

I did not assume that the thinking process would stop there, and I endeavoured to encourage it further by prescribing further reading, setting an essay and promising further discussion at a later date.

Having subsequently read some other writing about development, and Jessie Watson's article 'On the teaching of value geography' (*Geography*, **62**(3), 198–204), I think I would be inclined to structure a similar lesson in the future on the following lines:

1 *Empathy*. Start with an issue where students have strongly-held views (the issue of aid to developing countries seems to fit this category).

2 *Identification*. Make students aware of the cognitive and affective elements of their values.
Cognitive elements = reasons for aid-giving, the rate of population growth in certain countries.
Affective elements = the belief that we in the developed world are 'sensible' about resource-use, etc.

3 *Case-analysis and comparison*. Have the students describe and categorise the factual basis for their own views, e.g.
differences about the desirability of *large* families; the benefits of aid to donor countries as well as to recipients.

4 *Self-analysis*. Have students relate their own value-system to the values of others. Does new knowlege tend to reinforce their own views or change them?

5 *Choice and justification*. Examine some of the problems exposed through the giving of aid, and consider whether or not these can be solved (e.g. consideration of the breakdown of traditional divisions of labour through the introduction of new food-processing techniques, consideration of rural population through the location of highly-capitalised, aid-financed industry in urban growth-pole areas).

6 *Evaluation*. Have students write up their own viewpoint in the light of the evidence produced, and analysis made, and the view formed.

XII More shops for Abingdon! – *a decision-making exercise*

Eleanor Rawling

Here, role-play activity is seen in one of its most colourful and extended forms; Eleanor Rawling presents a major unit of work in which students are asked to play the parts of planning consultants in a local district. Data is provided and a decision required on the basis of the evidence.

This kind of activity has been part of the Schools Council 16–19 Project's encouragement of 'enquiry learning strategies'. What is described here in terms of a classroom exercise is also, in fact, an example of a decision-making exercise which can be used as an examination test.

In the new A level which the Project has established through the London Board, a decision-making paper, carrying 50 per cent of the marks of the total examination, will present problems of this style and level.

Introduction

Look at pages 69 and 70.
What would the 16- to 19-year-old geography student make of these instructions, presented in a geography lesson?!

He or she would no doubt be surprised, amazed, a little baffled and almost certainly interested to find out more and to attempt this exercise. After all, most of us are attracted by a challenge!

The challenge presented in this case to 16- to 19-year-old geography students is that of a decision-making exercise concerning the location of new shopping development in the small market town of Abingdon in Oxfordshire. The exercise has been prepared by the Schools Council Curriculum Development Project, *Geography 16–19*, as an example of a teaching approach which will help to train young people in the skills of enquiry, and, in particular, of decision-making. The project understands 'decision-making' to apply to a whole range of skills required by an individual in handling a question or problem and working through to reach a final answer or solution, capable of implementation. *Geography 16–19* considers that the ability to make reasoned decisions is an essential requirement of modern life, that 16- to 19-year-olds should be given every opportunity to practise this in school and that geography provides one valuable medium through which this can be done. In order to help teachers to plan effective teaching materials and approaches, the Project has put forward a route for enquiry. This route (see Fig. 15) provides a structure for

GEOGRAPHY 16–19: A ROUTE FOR ENQUIRY

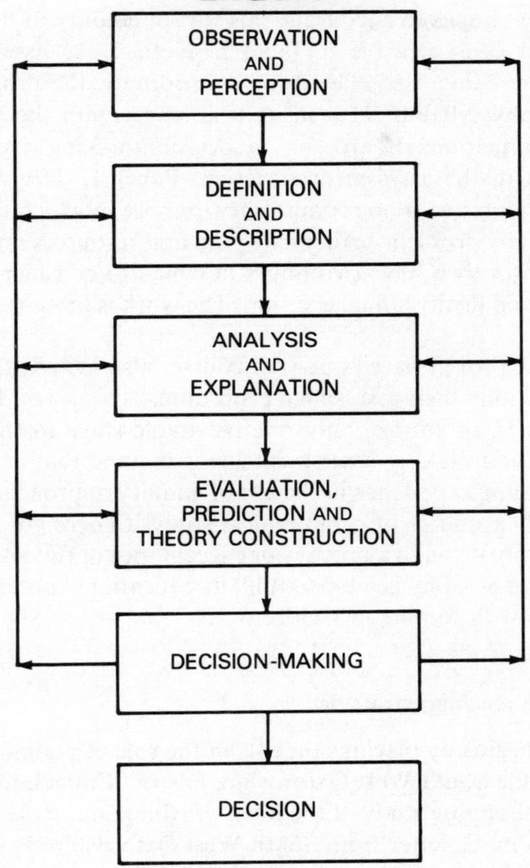

GEOGRAPHICAL ENQUIRY ENCOMPASSES NOT ONLY THE HANDLING AND ANALYSIS OF 'HARD' GEOGRAPHICAL DATA BUT ALSO THE ANALYSIS AND CLARIFICATION OF ATTITUDES AND VALUES RELEVANT TO THE PROBLEM OR QUESTION UNDER CONSIDERATION.

Fig. 15 *Geography 16–19: A route for enquiry*

the design of a variety of enquiry-based exercises. 'More Shops for Abingdon' is structured closely around this route.

If the Project hopes to encourage this sort of enquiry in the classroom, then it would seem sensible to build in methods to assess that these approaches are being used effectively. Accordingly, the Project's new A-level Geography syllabus, now in its trial stages with the University of London Examinations Board, has a decision-making exercise as one main element of the assessment structure. Paper 1, 'Decision-Making', presents candidates with one compulsory exercise in which they are faced with a case-study problem.[1] Adequate data and resources are made available. Candidates then have two hours in which to consider the evidence and to make and justify a final decision. The work is presented in the form of a report.

Many of the project-based one-year courses also include decision-making exercises within their assessment procedures. The use of these exercises in assessment is, of course, only one inevitable stage in the curriculum process. Decision-making assessment items assume that candidates will have been gaining experience in the use of enquiry approaches to learning throughout their course of geographical study. *Geography 16–19* is convinced that, only through a much wider acceptance of this sort of enquiry-learning, will geography be able to fulfil its educational potential and help to provide us with tomorrow's citizens.

Materials and teaching strategies

The exercise begins by placing students in the role of planning consultant requested by the South-West Oxfordshire District Council to carry out an independent shopping study of the town of Abingdon. Role identification is helped by a mock letter from South-West Oxfordshire District Council commissioning the consultant (see Fig. 16). The letter sets out the problem and the task. Abingdon requires a further 8 000 square metres of shopping floor space. Three sites have been found to be feasible for this scale of development (Fig. 17). The choice of each would result in distinctive impacts on the local area. A decision must be made as to where new shopping should be located.

The role-play element seems to be a good way of easing students quickly into the exercise, providing enjoyment and motivation. It must not be assumed, however, that decision-making exercises need always incorporate a role-play element. Indeed, it is often desirable for the student to be asked to justify his/her own personal decision, after having weighed up a variety of other possible viewpoints.

All necessary background information is provided for the students. This comprises the map of the town and three possible sites as in Fig. 17; some background information about Abingdon; the results of a shopping

SWOX SOUTH-WEST OXFORDSHIRE
Town Hall, High Street, Abingdon.

Fitzwilliam, Parker Associates, June 1979
Planning and Development Consultants,
St. Giles Chambers,
Grays Inn Court,
London.
Ref: CE/ABG/221/2

Dear Sir,

Shopping Study of Abingdon, Oxfordshire

The South-West Oxfordshire District Council has instructed me to authorise your firm to carry out an independent shopping study of Abingdon.

It has recently become apparent that:-

(a) Abingdon requires a further 8 000m^2 of shopping floor-space in order to continue its present role as a shopping centre.

(b) Three sites (A,B and C) are feasible for this scale of development in the Abingdon area.

(c) Future shopping development for Abingdon could take the form of one major development of 8 000m^2 at one of these sites or two smaller developments of 4 000m^2 each at two of these sites.

The District Council requires your firm to make a balanced assessment of the advantages and disadvantages (economic, environmental and social) of each size of development at each of these proposed sites, and on this basis make a

/cont......

Fig. 16 *Mock letter from South-west Oxfordshire District Council*

recommendation to my Council. Your final report should include an analysis of the environmental impacts likely to result from the alternative developments, a decision as to where new shopping development should be located (which site or sites of the three identified) and a reasoned justification for the decision made, showing how balanced consideration has been given to all relevant factors.

All necessary background information and data is available with this letter, including the results of an economic assessment of Abingdon's future shopping requirements carried out recently for my Council by Town and Country Property Surveys Ltd., and various letters of representation received from the public by the Planning Authority.

I await your report with interest.

Yours faithfully,

N. Smythe-Brown
Chief Executive Officer,
South-West Oxfordshire District Council

Enclosures:-

i. Fact sheet on Abingdon (from tourist brochure)
ii. Map at a scale of 1:50000, showing location of Abingdon.
iii. Map showing location of proposed sites for development.
iv. Detailed maps of sites for development at a scale of approx. 1:5000.
v. Summary of findings of shopping survey.
vi. Four letters of representation.

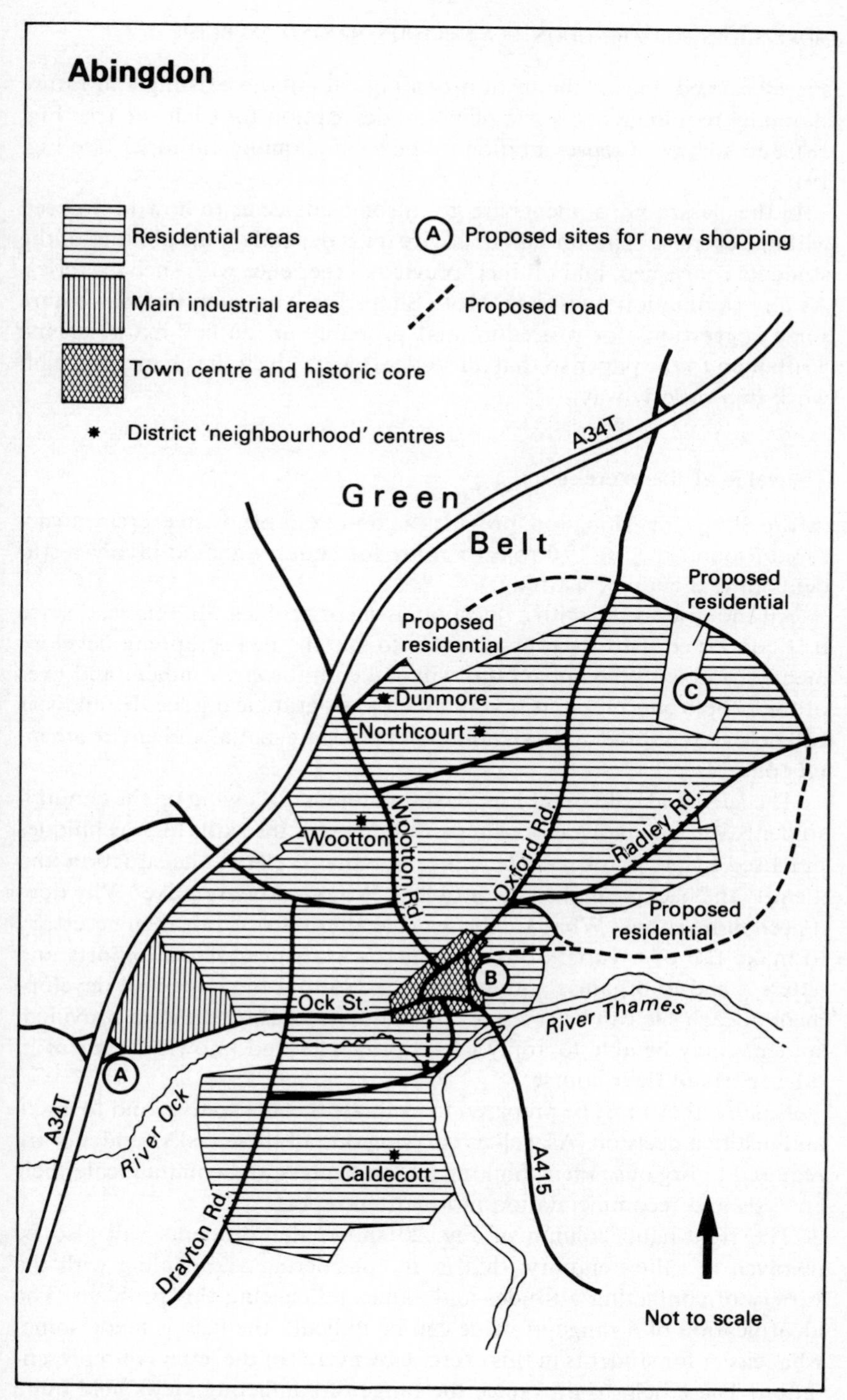

Fig. 17 *Three feasible sites*

survey carried out for the town providing data about existing and future shopping requirements; a site plan and description for each site (see Fig. 18); and a letter of representation to the local planning authority (see Fig. 19).

In the classroom, students are given some advice as to how to proceed with the exercise. The amount of advice must depend on the abilities of the students concerned, and on their previous experience with such exercises. As an examination question, 'More Shops for Abingdon' would require some suggestions for procedure and probably an outline mark scheme written on to the paper so that all students were given the chance to begin work in a orderly way.

The value of the exercise

'More Shops for Abingdon' provides a good example of an exercise structured around the 16–19 Project's route for enquiry and so involves students in real enquiry learning.

All enquiries begin with a question, issue or problem. In this case, since it is concerned with location – where to site the new shopping development? – and with the impact this will make on the environment and lives of the people concerned, it is very much a geographical issue. Its analysis will help to reinforce understanding of important spatial and environmental concepts.

The left-hand column of Fig. 20 shows that in following up the enquiry, students will be involved in using a full range of the skills and techniques practised by geographers. They have initially to clarify the situation and identify the background to the problem. What is Abindon like? Why does it need more shops? What are the feasible alternatives? It is then necessary to make use of a variety of data – maps, statistics, official reports and letters – and to organise and analyse this, so that the impact of development at each site can be assessed. In addition, to using the data provided, students may be able to apply useful principles and knowledge learnt in other parts of their course.

Finally, they must be prepared to evaluate the alternatives and to make and justify a decision. As well as carrying out all these tasks, students are required to organise their thinking and to express and communicate their analysis and recommendation in a written report.

The right-hand column of Fig. 20 shows that students will also be involved in values enquiry; that is, in considering and dealing with the variety of conflicting attitudes and values influencing this problem. The identification of a range of value can be difficult; the task is made somewhat easier for students in this exercise by means of the letters of representation. These help to introduce the range of conflicting views held quite reasonably by different kinds and different groups of people. It gives

Site A Marcham Road Approx scale 1: 5000

Marcham Road South

This site has the attributes of a major development site similar to the land already developed for industrial and warehouse uses to the north of Marcham Road. But because of its location away from the main residential growth area of the town, it is not the most appropriate location for local services and employment. It may be more appropriate, especially in the light of its accessibility to the strategic road network, to see it as accommodating some use fulfilling a regional role.

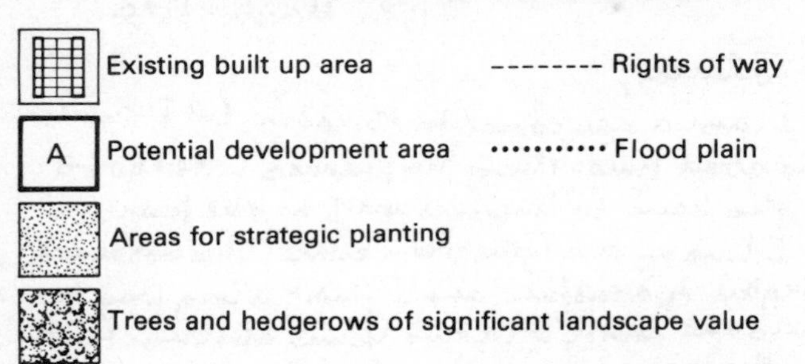

▲▲▲▲ Significant views into the area of potential change

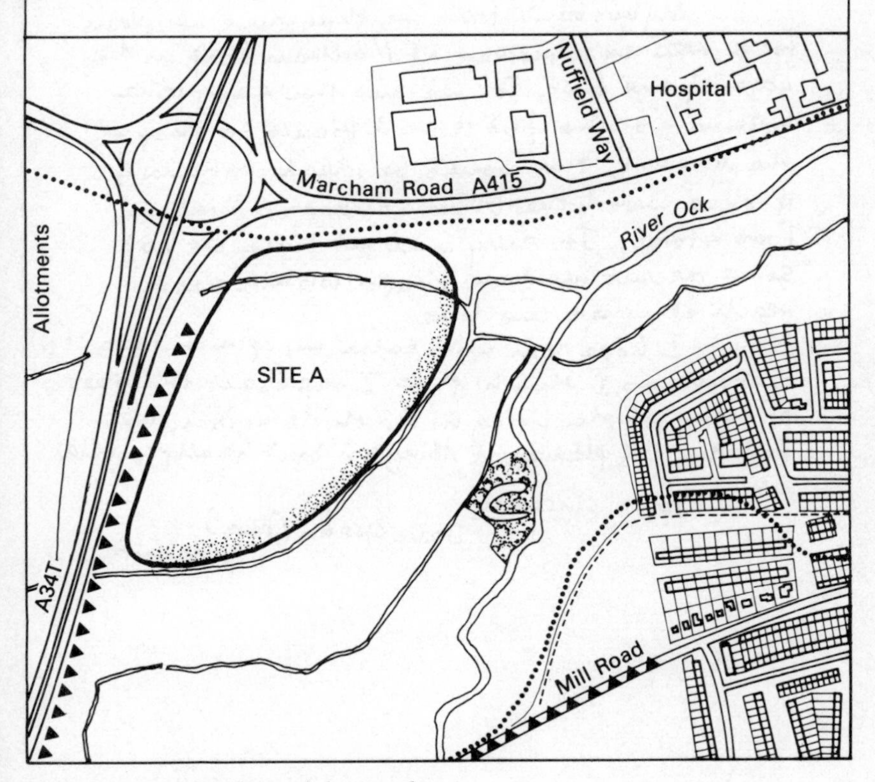

Fig. 18 *Example of one of the site plans*

43, Wordsworth Drive,
Tythe Farm Estate,
Caldecott,
South Abingdon.
15th March, 1979.

Dear Mr. Johnson,

I am a newcomer to Abingdon, but I think that it is about time that the planners listened to people who have to live and work in the town.

I live on the Tythe Farm estate. The bus service into Abingdon is dreadful, and I find it a long way to walk into the centre with two young children to look after at the same time. There is anyway, not a good food store with plenty of choice in Abingdon centre.

My husband tells me that there are plans for a new supermarket at Marcham Road on the edge of the town. For me and the many other housewives who find it so difficult to shop at the moment, this would be ideal, particularly if there were other kinds of shops as well as food stores. It really is about time we had some decent modern shops within easy reach of where we live.

I hope you will take my opinion into account as I think that I represent the views of many people living in South Abingdon. We feel that the planners of Abingdon don't really consider us,

Yours sincerely,
Jane Green (Mrs)
(Housewife)

Fig. 19 *A letter of representation*

	Stage	
Achieve awareness of the problem – need for more shopping development in Abingdon – which of three possible sites and what impact?	**Observation and Perception**	Achieve awareness that individuals and groups of people in Abingdon hold different views about the best site for new shopping development – from letters and background information.
Study background information about Abingdon and shopping needs. Look at data for different sites and different characteristics of sites. Consider information about likely impacts of development. Make preliminary hypotheses and consider how data can be organised to test.	**Definition and Description**	Consider for each site the likely views and opinions held by the groups and individuals represented in the letters. If possible, extend this consideration more widely to other groups of people in the Abingdon area.
Organise data into advantages/disadvantages of different sites and different sizes of development. Analyse the impact of shopping development on each site in turn. Accept/reject initial hypotheses. Consider any unexpected impacts/effects.	**Analysis and explanation**	Analyse how far people's views relate to where they live and to the impact the supermarket will have on their daily lives and travel (e.g. housewife). How far are their more personally motivated values related to abstract opinions (e.g. conservation – Friends of Abingdon). How can these be ranked?
Evaluate the different site alternatives in relation to present and future situation. Consider the implications of suggested strategies for one or other site.	**Evaluation and Prediction theorising**	Consider how an attempt might be made to rank values according to some scale of community need. How might your own views influence this? What might be the resulting preferred values and preferred site?
Decide which site has the desired capacity, physical attributes, convenience etc. and relate this to the values held/impact on Abingdon people. Make ...	**Decision making**	Relate these considerations to the factual enquiry which has shown which site is feasible. On this basis make ...
	Final recommendation	

Fig. 20 *A worked example of the route for enquiry in action*

students a way in to the complex task of clarifying just who gains and who loses as a result of development at different sites. Whilst the average candidate may, quite effectively, use merely the letters given, the more able student may realise that these are representative of a wider range of different viewpoints, and may be able to generalise more broadly about the values concerned.

Thus the final decision reached depends on clarification of not only what is feasible from a technical point of view but also what seems to be acceptable and beneficial to the people concerned.

In the Project's view, the education of students who will be required to make an effective contribution to modern society cannot afford to neglect this important element of training in values enquiry.

Whilst it is relatively straightforward to outline the intended benefits accruing to students who are involved in exercises such as 'More Shops for Abingdon', it is more difficult to find a sensible method of evaluating whether these objectives have indeed been achieved. Ultimately, of course, the test will be in the quality of students turned out by our 16–19 education system. More directly, the project's A-level Paper 1, 'Decision-Making', is to be marked by examiners. It is intended that assessment of the students' efforts will take into account the skills revealed in working through the question of problem, that is, the *process* of decision making. Although it is important that students are encouraged to make sound decisions, it will not normally be possible to assess actual decisions since most problems have several solutions! Mark schemes for decision-making exercises will reflect this bias. An example mark scheme for the 'Abingdon' exercise is given in Fig. 21.

Finally, it is worth pointing out that this exercise is based on a real

Candidates should note that marks will be allocated to the final report in the following way:

		Marks
i.	'setting the scene' – analysis of the background information about the existing situation and Abingdon's future shopping needs.	7
ii.	'identifying the alternatives' – finding out what the possibilities are for future shopping development.	8
iii.	'analysing the alternatives' – considering the impact that development at each site will make on commerce and business, on the immediate surroundings and on the lives of local people.	15
iv.	'weighing up the alternatives' – attempting to balance the advantages and disadvantages of each site and to make a final decision.	10
v.	'recommending and justifying' – recommending where new development should take place and providing well-reasoned justification for this.	10
		50

Fig. 21 *Suggested mark scheme for the Abingdon exercise*

issue. The people of Abingdon, the local Oxfordshire planners and indeed the Secretary of State for the Environment through the public enquiry system, are at present involved in a problem similar to the one presented to students.

It is to be hoped that sound decision-making skills are being exercised in Abingdon! – but countless other examples could be taken from real life. There can be no doubt that skills of enquiry and decision-making as practised in this exercise are essential in our modern world.

Note

1 The Abingdon decision-making exercise forms Paper 1 of the specimen papers for the Schools Council Geography 16–19 Project A level, mounted by the University of London Schools Examination Board. Figs. 15, 16, 17, 18 and 19 are taken from that paper, with kind permission of the Board and the Schools Council.

Part Two

Appraisals of the 1970s

Introduction

This section of the text includes some viewpoints and reflections about the changes in geographical education which characterised the 1970s.

Three schoolteachers - each of whom has remained in the same school throughout the decade - offer personal comment about their own experiences in the last ten years.

Sheila Jones had outlined the pattern of her lower school course at the 1970 Charney Manor conference and makes some comparisons in the light of its present state; John Rolfe's view comes from a school which has produced a stream of influential course books in the last decade and which has been a hearth of innovative classroom ideas; Pat Cleverley has taught in a large comprehensive school on a Bristol housing estate where the difficulties of resource constraint and environment have been a hurdle to climb in instigating development.

Rolfe voices unhappiness about the present place of physical geography in the curriculum, and also produced evidence of a surprising backlash against individualised work in the sixth form; his reservations about this are not shared by either Jones or Cleverley.

In view of the many misrepresentations of the 'quantitative revolution' at both higher education level and at that of secondary education, it is helpful to have a balanced note from Roger Robinson to put the record straight. Robinson has been involved with the Birmingham Teachers Workshop and with the Computer Assisted Learning in Upper School Geography (CALUSG) project in recent years, and he has done much to explain quantitative approaches to teachers and establish them in school work, without overpressing their claims.

Chris Joseph, himself deeply involved in examining work, muses about the general state of public external examinations, and points to some of the considerable changes which have taken place in the last decade. His general view of their present acceptability is cautious, however, and he voices a concern that they may be locked too firmly into a narrow mould.

Vincent Tidswell's paper is based on a major survey of 375 first-year undergraduates from nine British universities, and reveals a certain amount of disquiet about the imperfections of their knowledge subsequent to their A-level work. He attributes this to some 'inappropriate rote-learning of processes and principles'.

Like Joseph, he wonders: 'Are syllabuses asking for too much; allowing little time for absorption and reflection, thus being counter-productive in the attributes which really matter?'

Kirby and Lambert sound a similar note of concern about the present situation, and assert that middle-ability pupils find considerable difficulty with parts of the sixth-form syllabus. They take issue with a recent, optimistic pronouncement about geographical education made by Patrick Bailey, the editor of *Teaching Geography*, and argue that 'all is not quite as rosy as [he] would suggest, that contemporary geography is not the only geography and that evolution is necessary'.

They also identify some of the significant trends in academic geography – relative decline in the importance of quantitative methods, a greater involvement in substantive issues, a realignment in looking towards other disciplines for help and insight – and the later section of their paper covers some of the ground taken up at greater depth by Derek Gregory in Part Three.

Finally, in this section, Richard Daugherty makes an important contribution in a survey of the 1970s which outlines the growing activity with the field of curriculum politics. Whereas, at the beginning of the decade, most geographers were rapt in *in*ternal discussion of the nature of a changing discipline, by the end of it, increasing pressure from outside was forcing a change of stance.

Daugherty points out this, despite the stiffening of geography's academic sinews, it played little significant part in the considerations of those who wrote the initial documents which presage a greater 'centralised' interest in the curriculum.

He suggests that it is vital for teachers to actively justify the relevance of their subject within the general aims of education and not to assume that it will be there by acknowledged right for ever. His paper sets the stage for much of what the next ten years may be about in geographical education.

1 Three schoolteachers reflect on geography 1970–1980

Sheila Jones, John Rolfe, Pat Cleverley

Sheila Jones Colston's Girls' School, Bristol

In recent years I have often heard pupils say 'I wouldn't want to teach; it must be so boring teaching the same thing year after year.' Charney 1980 forced me to consider what has happened to geography at Colston's Girls' School in the past ten years and I now feel that in the future I shall have a strong counter argument to this statement.

Since 1970 Colston's has faced no major organisational changes and so we have been lucky enough to be able to concentrate on teaching and curriculum change. The overall aims of the department are as in 1970 but changes in the external examinations we take have resulted in alterations in the Ordinary and Advanced Level courses. The course for the first three years remains almost unchanged from that described in *New Directions in Geography Teaching* (pp. 144–7).

However, there has been a major development at the start of the first year consequent upon a decreased knowledge of basic mathematics and geography by the eleven plus entry. It had become increasingly difficult to begin the geography course with plan and mapwork because of the inability of many girls to cope with the concept of scale. At the same time the mathematics department was finding that the diversity of knowledge and understanding within a three-form entry from a possible seventy schools was considerable, and that some problems could lie dormant for several terms. So was born a foundation course in mathematics and geography, with a pupils' booklet produced in department co-operation and shared teaching for half a term, with staff from both departments working through such topics as fractions, decimals, scale and co-ordinates. The mathematics department can now begin its remedial work at a very early stage and the geography department gains by knowing exactly which mathematical skills the pupils possess and how these have been presented to them.

The greatest changes have taken place in the Ordinary Level course, for Colston's was one of the eleven pilot schools participating in the Schools Council 14–18 Project. Since 1974 seven groups of candidates have taken this examination with a pleasing success rate. It was not felt that SUJB Advanced Level was a natural follow up to this course and so in 1974 we started work for the Cambridge examination. If the change were to be made in 1980 the choice would be less straightforward and the new 16–19 Advanced Level administered by the London Board would seem to be a more natural continuation....

The changes in the factual content of both Ordinary and Advanced Level courses have been considerable and in the former the whole ethos of the examination has affected teaching and learning methods. The 14–18 Project has been well documented and the detailed examination regulations and a study of past examination papers give a clear outline of its main features. The change in content on the Advanced Level course has been almost as dramatic as that at Ordinary Level, particularly in physical geography where there has been a change in approach (a concentration of processes) and the inclusion of new topics such as plate tectonics, almost unheard of in 1970. The content of human geography too has changed, but many of the topics now included were already well established in more specialised and general sixth-form work in 1970.

Throughout the school one major development which has accelerated and aided changes in both content and approach has been the proliferation of textbooks, particularly in the past five years. Ten years ago was the time of the 'underground press' when every teachers' meeting, course and conference produced a welcome hand-out exercise. The diffusion of ideas at that time could well be related to the travels of Rex Walford, Geoff Dinkele and others who generated the prototype schemes of work which others copied. Careful planning and organisation of departmental funds in the past ten years has meant that we have built up a store of stimulating and useful books, many of which can be used at more than one stage of study.

A further change in course content has resulted from an increased ability to handle statistical data because of the widespread use of pocket calculators now owned by most of our enthusiastic Ordinary and Advanced Level candidates. The immediate future promises further changes as a school micro-processor is to be installed in the autumn of 1980.

How have such changes affected departmental organisation? In a small school communication has always been easy and there has been considerable continuity of staffing. Thus ideas and innovations have been easily disseminated. The Ordinary Level course has both strengthened and improved relationships, for the allocation of a time-tabled period for department meetings together with team teaching and internal moderation have assisted the inexperienced and re-educated the old hand. Innovation work, essential for the 14–18 Project, has extended throughout the school and the research work, discussion and preparation of materials are now more carefully organised and structured. Teaching styles have altered little, for the idea of the teacher as a change agent within a classroom situation had already been developed in the lower forms, but can now be used at the

Ordinary Level stage. However, that is just one approach, for it is felt that there is a place for a variety of methods within the teaching situation.

There have been some observable changes in external influences. Within the Staff Common Room, after some initial conflicts of interest because of the demands on pupils' time in the OL course, there is now an appreciation rather than a criticism of the changes and a positive and helpful interest in fifth and sixth form Individual Studies. Inter-departmental co-operation has increased, including work with the physics department on sixth form slope exercises. There is some criticism from colleagues of the lack of locational knowledge, especially from the linguists and historians. This seems part of the general malaise causing anxiety concerning spelling, arithmetic and grammar both in the staff room and amongst parents. Certainly, in spite of departmental efforts to remedy the situation, the syllabus which we follow does militate against the 'capes and bays' factual knowledge and to many of the general public geography teachers may not seem to be considering the basis of the subject. However, it is heartening that in our school situation, the general consensus amongst the parents seems to be that the geographical knowledge being gained by their daughters is more relevant and more stimulating than that of their own schooldays. We are also lucky that the majority of parents are very supportive with field work and are both able and willing to act as chauffeurs, surveyors' mates and general dogsbodies.

In the past ten years the external influences of the media have also increased. Colour television is now widespread and video recordings make it possible to time-table programmes to the department's convenience. BBC radio programmes under Geoffrey Sherlock's guidance have kept up with (if not ahead of) developments in our subject, which no self-respecting teacher of geography can ignore. Specialist publications include the stimulating *Geographical Magazine* (now available to sixth form and staff at reduced rates), the admirable and entertaining *Geo* for Ordinary Level candidates and the Geographical Association's *Teaching Geography* for the staff.

Lastly, what of the customer? It seems to me that in the first three years of secondary school there have been few major changes in pupil attitude, although school is much less the centre of the universe than it was ten years ago. Ordinary Level candidates now have to work as hard as ever although in a different way. Hard work for two years cannot guarantee a C grade, but the chance of success for the average candidate is now greater. Not only can a good Ordinary Level grade be achieved, but also the personal satisfaction of the presentation of a piece of individual research. The numbers choosing to take geography at Ordinary Level have changed little although increasingly the more able scientists seem to settle for geography whilst those more committed to arts subjects select history; the latter is also chosen by those who prefer the safer option of a clearly defined syllabus in which a basic factual knowledge can still bring success....

It is at Advanced Level and in the post-Advanced Level choices that there have been most changes. A greater career consciousness and desire for vocational training have been to the detriment of the subject in the sixth, with economics now a strong alternative and there is a loss to the pure sciences of many able girls. As a department we are not always successful in convincing some of our more promising pupils that a very good geographer might in the long term have better opportunities than a weak scientist. Fewer girls are selecting geography as a specialist subject for further education courses, but this is also true of a number of other subjects. Current publicity and parent pressure are encouraging them to courses in languages, business studies and science. Hopefully, when the economic climate improves, pressures upon career choices may decrease, to the advantage of geography.

Changes there have been – some, as in the 14–18 Ordinary Level, radical – I hope that the department has never changed for change's sake, but only following a careful assessment of the consequences to the department, to the curriculum as a whole and to the pupils. By 1990 I shall have retired and no doubt my successor will be introducing the first of her changes in the evolutionary development of the geography department at Colston's.

References

Textbooks in Colston's Girls' School

Middle School

RICE, W. F. (1975, 1978) *Patterns in Geography*, Books 2 and 3. Longman

Ordinary Level

ATHERTON, M. and ROBINSON, R. (1980) *Water at Work*. Hodder and Stoughton

DUNLOP, STEWART (ed.) (1976) *Place and People*, Book 1. Heinemann

JONES, MELVYN (1979) *Assignment Geography*. Hulton

SAUVAIN, P. (1972) *Advanced Techniques and Statistics*. Hulton

Advanced Level

BRIGGS, K., RILEY, D. and TOLLEY, H. (1979) *Data Response Exercises*. OUP

ELKINS, T. and CLAYTON, K. (eds) (1974) *Aspects of Geography* (series) (e.g. 'Slope Development', Young, A. and D.), Macmillan

GUEST, A. (1974) *Man and Landscape*. Heinemann

HILTON, K. (1979) *Process and Pattern in Physical Geography*. UTP

MONEY, D. C. 1972) *Patterns of Settlement*. Evans (also others by the same author)

ROBINSON, R. (1978) *Data in Geography*. Longman
SMITH, D. and STOPP, P. (1978) *The River Basin*. CUP
TIDSWELL, W. V. (1976) *Pattern and Process in Human Geography*. UTP
WEYMAN, D. and V. (1976) *Landscape Processes*. Allen and Unwin
WYNNE-HAMMOND, C. (1979) *Elements of Human Geography*. Allen and Unwin

Further reading
Examination regulations for the 14–18 Geography Project; and Question papers 1977–9 inclusive
NAISH, M. (1980) 'A Medium for Education', *TES*, 18 April
WALFORD, R. (1973) *New Directions in Geography Teaching*, pp. 144–7. Longman

John Rolfe The Haberdashers' Aske's School for Boys, Elstree

Since the appearance of *New Directions in Geography Teaching* in 1973 it cannot be said that Charney Manor meetings have enjoyed any monopoly of reflection on the state of the discipline in schools. Contributions have come thick and fast, notably from the Inspectorate,[1] David Hall[2], Michael Walker,[3] four Schools Council Project teams, W. Marsden,[4] Norman Graves,[5, 6] M. Williams[7] and many others. To the teacher who from time to time attempts to step back for a moment and perhaps brush the dust from his teaching notes Dr Graves declares reassuringly, 'It all seems relatively simple ... the teacher's curriculum problem appears to resolve itself to finding suitable objectives, arranging appropriate learning experiences for the students and to finding ways of evaluating these learning experiences.'

Taking the first task, objectives are probably the areas where most classroom teachers are least confident. Indeed, only recently Keith Orrell, Director of the Geography 14–18 Project, stated that he had received a coursework assessment synopsis from a Project School and under the heading of objectives the teacher had written – 'Would you be good enough to fill this one in for me?' It is my impression that school curriculum planning at such comparatively higher levels as the N & F deliberations and Schools Councils Projects have suffered similar hang-ups in their initial stages and appeared likely to be doomed by spending too much time on the first question on the paper. This is not to deny that objectives should be the one compulsory question on the paper and not like so many candidates' essays where the introduction falls into place when the rest of the essay has already been written. But the teacher has been on much firmer ground with the remaining questions of subject content, teaching strategies and evaluation techniques. All three were given a magnificent lead in the early 1970s with the publication of the American High Schools Geography Project. At that time £1m. had already been invested in the Project yet it was estimated that by 1976 penetration and presumably take-up throughout the United States was to a mere 5 per cent of social studies teachers – hardly an encouragement to cost-benefit minded publishers in this country. Significantly, teacher-produced materials have proved popular in the Middle School range – those published by Arnold, Harrap, Longman, Blackwell and OUP spring to mind! Even at the sixth-form level students turn to books like Vincent Tidswell's, Bradford and Kent, Meyer and Huggett, John Bale, Darryl and Valerie Weyman, Everson and FitzGerald and others whose roots are or were firmly in the classroom. To my mind this phenomenon is surely one of the pluses of the decade and shows a great contrast with the limited choice of publishing materials aimed at the classroom a decade ago when the Honeybone series or Young and Lowry ruled supreme. And it could have been the dearth of suitable material or the combination of that and the introduction of integrated studies and teaching mixed ability groups which acted as the springboard for so much of the teacher-produced work cards which have become the hallmark of teaching in the 1970s.

Today, with the aid of such technical aids as the Banda machine, the overhead projector, the photocopier, the electric stencil cutter and offset-litho machine, whose results are sometimes seen in colour, the teacher has emerged as a most effective innovator in producing his or her very own original teaching units, best suited to the needs of a particular, far from homogeneous, set of pupils. (The opportunities thrown up by computers and classroom technology is taken further by my colleague Michael Day elsewhere in this book (see 'Towards a new generation of teacher-technologists?').) Yet, whenever I go to the GA Annual Conference or other exhibition of recently published books to search out such gems I am told by geography editors and representatives that geography teachers who spend so much of their time originating such teaching aids are rarely the ones who submit their work to publishers. A pity!

In terms of syllabus content the 1970s are surely best remembered for the demise of regional geography in the schools – in the old sense at least. Few examination syllabuses for 16- to 18-year-olds today expect much in the way of global coverage but during the past two or three years impassioned pleas for a return to 'the real world of regional geography' eschewing the imaginary world of models and key concepts have been made, notably from the directions of Bristol and the London A level examiners. Recently the importance of area studies has been reiterated by London, emphasising that they should be studied in conjunction with the six basic systematic topics and they have gone to considerable lengths to suggest suitable topics.

The role of physical geography remains an enigma to most of us. In the minds of some teachers physical geography should be firmly placed on that side of the fence which includes the other earth sciences. Others argue equally cogently that hydrology and biogeography are as integral to the discipline as resource use and conservation, pollution and planning at various scales might be. Such a fundamental argument was given considerable thought and hot debate during some of the N & F deliberations. In the end it was decided by all three Schools Council groups to retain physical geography as an integral part of post O-level studies. Following the rejection of N & F one of the most encouraging decisions of the 16–19 Project Team was that the geography curriculum should be entirely orientated around man–environment questions, issues and problems. The plea for greater relevance to the needs of the younger generation may be answered with this shift of emphasis. If our own student reaction is anything to go by, sixth form geography would be much more popular if it contained less pure meteorology and geomorphology and more of an applied geographical nature with a stronger regional emphasis than of late. Most syllabuses are hopelessly overloaded anyway and one must be selective in one's approach. There simply isn't time to spend six weeks introducing hydrology and then be faced with a highly specific question on an A level paper twelve months later.

The subject at six form level, at least, is perhaps in a greater state of flux than ten years ago. At that time models and quantitative techniques were clearly marked curriculum chunks for the geography teacher. These were readily filtered down from university department to the sixth form geography room. Looking ahead to the 1980s can one say the same about a welfare approach, let alone a radical one? Perception has never really got off the ground; likewise the plea for political education within the discipline has hardly been welcomed with open arms. Somewhat sadly it might be said that innovation in the content of geography in schools has fizzled out at present and teachers are looking for a lead.

In the classroom the teacher is on much firmer ground. An occasionally noisy classroom today does not raise quite so many head teachers' eyebrows as it did in 1970 as games and simulations have become almost respectable. Indeed, decision-making, enquiry-based learning, a conscious effort to 'teach attitudes' and enable students to acquire communication skills, thinking skills, social and practical skills as well as study skills have all been added to the teacher's repertoire. The once all-conquering models have now become the *bêtes noires* of teachers and examiners at all levels. Even field work/field research finds itself in something of a rut.

In the sixth form we alternate a field week of a CBD study of a city, the local farming system, industrial location and planning with a following year's studies of stream measurement, valley-scale meteorology, a Reilly and Huff exercise and perhaps the conflict of recreation and industrial interests in a National Park. Important though these might be we have pursued such work for ten years and would love to see some other suggestions. Of course, cost to the students is the overriding factor here. At ten to fifteen pounds a day per student, one has to be very sure that what you are doing is essential, and the ten-day excursion to Italy of fifteen years ago has now shrunk to three nights in Brighton.

Evaluation has provided some of the brightest aspects of the decade. In the independent section Ellis-inspired curriculum design meetings for 9- to 16-year-old pupils in preparatory schools and beyond have finally broken the Common Entrance Examination impasse. Or was it the other way about?

For us, apart from our involvement in the Oxford Geography Project books, undoubtedly the most rewarding experience in the Middle School has been the 14–18 Project. Even as a pilot school brought in as something of an outsider and largely unsupported by the Project, the effect on staff and pupil alike has been astonishing. To date nearly 800 candidates have produced individual studies and by careful permutation the titles have yet to be quite identical. Teachers have spent days composing questions for the Chief Examiner, marked ninety internally set and assessed coursework assignments over half term, read candidates' individual studies during Christmas holidays without pay (until this year), but in return the Project Team and Cambridge Syndicate have produced superb Ordinary Level papers. In 1979 a total of 4 700 candidates took the summer examination and 67 per cent obtained at least a Grade C. Perhaps of even wider significance is that 16–19 GYSL and 8–13 Projects have agreed to combine in seeking to initiate regional centres for geographical curriculum development. A common O level/CSE, pilot 16+ examination on 14–18 Project lines could undoubtedly lead to a massive take-up. A plan from a sympathetic government willing to introduce a 16+ examination of this kind within the next decade would meet with wide approval in schools and industry alike.

For potential sixth-form students *Geography 14–18* has presented its own problems. Of our ninety-three O level candidates in 1980, forty have started to take geography at A level but their ultimate take-up may largely depend on three conditions being satisfied. First, that further individual studies are *ex*cluded; second, that the amount of pure physical geography will be drastically reduced; and third, that internal assessment is dispensed with. A case of the tail wagging the dog or proper consumer-orientated change? (We now take the London A level examination, having changed from Oxford and Cambridge Examination Board.)

Q & F closely followed by N & F have presented us with a much needed opportunity to throw everything into the melting pot for the 16–19 student. The three N & F geography schemes all had their own merits and no doubt new A level syllabuses will benefit from them. From a personal viewpoint the most disappointing aspect of the rejection of N & F and the consequent loss of the possibility of sixth formers taking up to five subjects at N level was the realisation that as a result geography might have suffered a blow at sixth form level in the long term. Careers staff advise that three science A levels will be the students' best chance of avoiding unemployment, and they may be right.

In the third year sixth the prospects of an Oxbridge place, let alone an award in geography, are becoming gloomy for young men. There is no doubt that it is much more difficult for a young man to gain admission to an Oxbridge college on the arts side than it was in 1970. At that time the way ahead seemed comparatively clear, but ten years on the classroom teacher feels the need to consolidate on what has taken place in the 1970s. I view the next ten years with much less certainty

References

1 HMI (1978) *The Teaching of Ideas in Geography*. HMSO
2 HALL, D. (1976) *Geography and the Geography Teacher*. Unwin Educational
3 WALKER, M. et al. (1977) *Expectations in the Field of Humanities*. Oxford County Council
4 MARSDEN, W. E. (1976) *Evaluating the Geography Curriculum*. Oliver & Boyd
5 GRAVES, N. J. (1975) *Geography in Education*. Heinemann Educational Books
6 GRAVES, N. J. (1979) *Curriculum Planning in Geography*. Heinemann Educational Books
7 WILLIAMS, M. (ed.) (1976) *Geography and the Integrated Curriculum*. Heinemann Educational Books

Further reading

BODEN, P. (1976) *Development in Geography Teaching*. Open Books
GRAVES, N. J. (ed.) (1972) *New Movements in the Study and Teaching of Geography*. Temple Smith

Pat Cleverley Withywood School, Bristol

When I transferred from the quiet glades of a grammar school to a comprehensive school on a council housing estate, one initial shock, amidst many, was finding that each student did not have a textbook and an atlas. Inadequately financed when opened, the geography department's sets of books were of twenty maximum, and could not therefore provide a basis for homework or revision.

Hence the duplicating revolution, which had begun with the multicoloured Banda, increased rapidly in the face of necessity. Within this decade we have tried a multiplicity of types, thinking each investment to be the last word in reprographics, only to be superseded by the next miracle machine. Large schools had the advantage of adequate financing for off-set lithos, and this has greatly improved the appearance of our worksheets. The consumer complaint of 'not another sheet' was partially overcome by binding facilities and topic booklets.

Other audio-visual aids and facilities have also improved greatly. Colour has proliferated, especially via slides and filmstrips, though cine films have not received much increased use. They are more likely to be spotted by students as historical rather than geographical, and presumably geographical education has not been adequately profitable for the film-makers. Television recording and replay has increased, although the age of several easily viewed colour sets per class has yet to reach us. The BBC schools department has increased its output in this period, and its quality is amongst the best. More would be appreciated, when finance permits.

The original complaint of the lack of a book per student in the long run may have proved advantageous. If work was going to be duplicated in any case, why not write your own? But those who have been involved in such curriculum change within their own schools will realise that this simple idea is liable to collapse into a series of pitfalls.

The speed of innovation has varied, and in the earlier years of the decade it was very patchy. Early efforts at innovation often took an isolated exercise or idea and injected it into a reasonably possible place in an inappropriate syllabus. This provided light relief for staff, and hopefully students, but did not have the impact of later developments. Exercises were acquired at meetings and courses, which were often evaluated by the number of hand-outs.

The difficulty was deciding where to start. Before *Geography 14–18* there was only one choice; at the bottom, and for most of us this was the first year of an 11–18 school. Re-writing the syllabus and course, even one year at a time, was a major undertaking if the number of committed staff was small. Were there any books suitable? Could we really abandon all those expensive tomes previously thought to be indispensable? Would we have endless complaints from the parents unable to do their children's homework? Would the students enjoy it as much, or more, than the previous course?

Before most of us got fully organised throughout the Lower School, *Geography 14–18* arrived. As one of the original trial schools it was as long ago as 1971 that we started writing the material leading to the first O level in 1974. This was the point at which we really appreciated the reprographics revolution and readily realised how difficult it was to devise 'original' material. It felt as if we were drowning in a sea of paper, not to mention educational jargon. Concepts were definitely in – and could it be true that facts were out?

Let off the hook of a packed O level syllabus, we opted for in-depth studies of a few topics. Often these were not even mentioned in the A level texts of the day. If the department was sufficiently large, two people could deal with each topic, research it, write the material, conjure up the assessment and be responsible for a never-failing supply of hand-outs.

The place of field work was reinforced, but its character needed to change. It was no longer adequate to walk and view and sketch the coastline: pebbles had to be measured, cliff angles calculated, vegetation plotted and scattergrams attempted. Urban studies involved perception as well as observation, and the well-drafted questionnaire, perhaps with portable cassette recorder to hand, replaced the gentle walk and talk.

Individual studies were another new departure. In the comprehensive school, with many students of average ability doing a mixture of O levels and CSE, the greatest difficulty was, and still is, to clarify the difference of the O level Individual Study from the CSE project. Many would be happiest with reading only as the basis, and field work entirely on one's own seems to present insurmountable barriers. Having overcome this handicap, the greatest difficulty remaining is finding enough different topics that can be attempted within a council housing estate. Bus fares are expensive, and in an area of high unemployment or single-parent families, for many the estate is the limit. Those interested in non-climatic physical geography have the greatest handicap.

Within the Comprehensive, O level alone will not suffice. But here the CSE examination has been very flexible. The same basic material as developed for the O level course formed a basis for a Mode 3 CSE, and has proved to be much more satisfactory than the Mode 1. Although able to modify the material for those immediately below the O level standard, we still need a different, and in some ways more traditional, course for the well below average. But even here the new approaches, material and attitudes have led to a separate Mode 3 course dovetailed to those parts of the main course found most satisfying.

At A level change has been slower. The universities have been more conservative and fearful of a lowering of standards, especially if that dreaded ogre 'teacher assessment' was suggested. Different boards have modified at their own pace, and it is unfortunate if after *14–18* O level, you have to study for a conservative A level. Unlike the O level, there has been no equivalent examination at A level, for which you can enter irrespective of your school's choice of examination board.

A-level modifications have shown some adaptation to the impetus from beneath, and individual examiners show appreciation of these trends in their questions. The syllabus does read a little better, but it is just as full as ever, and the papers are usually arranged so that you have to cover the majority of topics. Allowing for the increase in level, the scope for development in-depth is less than at O level.

However, change is about to happen here as well through the work of the 16–19 Project, and in the next decade we may see developments as beneficial at this level as at 16+. I hope that these will include courses more suitable for the one-year sixth former.

After the innovatory 1970s the pleasures of consolidation are fortunately now with us, and we are even girding our loins for the necessary renovation or the next revolution. I hope that in 1990 I am still as committed to gradual change and evolution as now, and I hope that changes are equally stimulating and beneficial.

---

2 Quantification and school geography – *a clarification*

Roger Robinson University of Birmingham

The 'new' human geography of the 1960s would have been impossible without quantification. It was indeed such an important feature of it that many people confused 'quantitative' geography with 'new' geography. In fact, the essence of the 'new' geography was more in the development of models and generalisations based on the patterns created by 'average economic man'. Quantification is a tool used not only by the school of empirical model builders, but also by the subsequent behavioural, radical and humanistic geographers. Almost all modern geographers are well versed in statistical methods and happy to use them to describe and analyse information.

In school geography when the term 'quantification' is used it usually refers to one or more of the following four main areas:

1 *Numerical description and measurement, presented as 'statistics'.* Information in geography has always been presented quantitatively, and measurement of production, population, climate, relief, etc., universally accepted. The problem faced in recent years is the abundance of such statistics, and the difficulty of selecting appropriate small sets of data for use in the classroom.

2 *Numerical analysis.* (a) To identify patterns and distributions. Graphing (line graphs, bar graphs, etc.) has been accepted as a technique worth teaching – especially as the traditional presentation of temperatures and rainfall patterns. Similar methods of representing spatial information – isopleth and choropleth maps – are often used but seldom constructed. Ranking, dispersion measurements (quartiles, etc.) are examples of simple techniques sometimes used in school.

(b) To identify relationships and associations.
The techniques easily used in school are mainly diagrammatic - e.g. scattergraphs, tracing overlays of maps - and quadrat samples and matrices are used at some advanced levels.

3 *The use of statistical indices.* The use of indices simply as descriptions of relationships or association between specific variables is quite acceptable and within the grasp of A level students. The 'nearest-neighbour' statistics, Spearman's Rank Correlation Coefficient, Pearson's Correlation Coefficient and Association Indices (perhaps Chi Square) are possibilities.

4 *The use of statistical tests and statistical inference* (e.g. chi-squared test, student t-test, etc.). It is unlikely that most A level geography students have the time to develop any real understanding of the use of tests and the inferences that may be drawn from samples, unless they are also studying mathematics or statistics.

In a school committed to quantification as an essential element of geographical study it is likely that area 1 above is present in the curriculum of every year group. Area 2 is developed and emphasised by years 4 and 5. Areas 3 and 4 are the province of A level sixth-formers.

Pupils, teachers and education 'experts' have offered all kinds of arguments for and against quantification in school geography. The table below juxtaposes some opinions often expressed.

Table 5 *Quantification in school geography*

For	*Against*
The understanding of simple statistics and the application of numerical skills is important for everyday life	Work with quantified information is too difficult for most students, and such an approach will create a barrier to learning
Quantification is an important dimension of description	It makes information boring
Patterns and associations are made clear	The numbers are often unrelated to reality and the 'patterns' questionable
Helps students to make informed decisions, and to take a more objective view of problems	Quantified information is often irrelevant to real problems, and focusses attention only on features that can be measured
Gives intellectual satisfaction because it permits a scientific approach to some problems	Measuring and calculations are unnecessary distractions

Table 5 *Quantification in school geography—(continued)*

For	*Against*
Gives extra meaning and new possibilities for field work and personal investigation	The quantitative work is too time consuming
Provides more opportunity for exploration and analysis by students	Becomes the dominant element of student activity
Forces teachers into a reappraisal of the content of their syllabuses	Number skill may be seen as the central skill in a geographical curriculum
The organisation of the ideas in a syllabus is improved, and more rigour is given to the study	Quantification is tagged onto existing syllabuses as an extra to please the 'new' establishment
Statistical indices provide excellent descriptions of patterns, associations and relationships	Statistical Tests and Inferential Statistics are badly taught, misunderstood and inappropriately used in schools

There is certainly some truth, sometimes, somewhere in all these statements, and it is useful to see how they apply in your own experience.

It seems that the quantifiers should take care not to mis-use or over-indulge their predeliction, and especially to beware of over-emphasising the 'economic man' model's philosophy that came with the first flush of quantification in human geography. Non-quantifiers are in an awkward position, since almost all academic geography accepts the use of quantification when appropriate. They need to justify their omission with this in mind. This does suggest that any course aimed at giving the students the chance to be 'geographers' (e.g. A level?) must include elements of quantification.

But the most significant implication of quantification is in the development of an attitude of mind where students are not afraid of numbers as an important means of description, and accept numerical analysis as one aspect of geographical study, and an approach to problem solving. This attitude of mind can be nurtured in the lower school without turning geography lessons into a mixture of maths and statistics. And this attitude need not overshadow other equally important attitudes. 'Quantity' is certainly no substitute for 'quality', but it is hard to imagine a credible school geography of the 1980s without numbers.

3 Examinations in the 1970s

Chris Joseph Marlborough College, Wiltshire

Any attempt to survey the whole field of geography examinations in the 1970s is over-bold. To speculate about the 1980s can only be rash. Most teachers though are heavily involved with one system and find it hard to distinguish the wood of progress from the trees of local problems. A summary is worthwhile if it leads to reflection and directs energy into profitable channels for reform.

Looking back to the early 1970s, the most striking features were the strength of regional geography and the simplicity of exams. Systematic studies were weakly developed and theoretical work almost unheard of. The sketch map and the written account were well to the fore:

> Select one of the following rivers. . . . With the aid of a sketch map describe the main relief and drainage features of the river basin chosen. In what ways has the river (i) helped, and (ii) hindered man's activities in the area?

'Explain why Wales is a very unevenly populated country' harks back a decade further to the time when the candidate was asked to: 'Write an orderly geographical account of North Wales'.

Such simplicity is music to some teachers' ears. But the marks from these questions tend towards a high mean score with a small standard deviation. Any grade discrimination based on them is likely to turn on much narrower margins than teachers might guess at or be comfortable with.

Today, much has changed. Diversity makes generalisation difficult. For most teachers the regional model is no longer the framework for organising their syllabus, nor is it much evident in the spate of new books each year. This change is reflected in the innovative structures of the Schools Council Bristol Project (14–18) and Geography for the Young School Leaver (GYSL). Both have abandoned regional patterns and a physical/human division in favour of new themes such as the city, with stress on enquiry-based learning. Formal examining continues, but there is a large role for the teacher in the final grading.

> Coursework, including individual studies (50%), is internally assessed and externally moderated with the schools determining the rank order of the candidates on the basis of their coursework. (*Schools Council 14–18 Project Handbook*)

These projects have rightly attracted widespread attention but it is only fair to see them in the context that, in 1980, the Bristol Project was taken by 3.5 per cent of O level geography candidates and GYSL's CSE by 11 per cent of the entry. Still, these figures increase yearly and the ideas and materials of the Projects spread, like ripples in a pond, far out from their original centres. The main limitations to widespread adoption may be money and teachers' time. Before me is the material set for an O level Paper I in 1980. It has: (i) a question paper of eight A4 pages of text and maps, (ii) a booklet of eleven pages including three oblique air photos and further diagrams, (iii) a 1 : 25 000 OS extract. It makes a superb resource collection, but the printing costs come high. Perhaps the sheer complexity will end such papers. Invigilating candidates, in any subject, shows a teacher that students have to face a maze of material on their desk, which can easily baffle them, and lead to failure through the dreaded 'rubric offences'.

The new projects require a formidable commitment from teachers in assessment and preparation, plus a wide variety of books, maps, slides and other resource material. Here teachers' centres are playing an increasing role in sharing costs and pooling both equipment and ideas. Their sustenance and further spread will be vital in the impoverished 1980s if new projects are to have the success they deserve.

Amongst the major GCE boards the prevailing fashion at O level is now for systematic studies of the British Isles, and a world problem or thematic approach to places elsewhere. This is often in the form of data response questions and these can be both precise and challenging.

> Describe and account for the main features of the planned lay-out of the New Town shown on Map II of Cumbernauld. Explain: (i) why New Towns have not always been as successful as was intended; (ii) the attraction of New Towns for certain industrial developments. . . .

There is a possibility, though, that such syllabuses leap-frog a decade of quantification straight to the welfare and behavioural approaches that are currently fashionable. This may lead to questions that severely tax the conceptual reach of candidates:

> 'The Green Revolution is now failing to keep pace with the rapid growth in population, and this will condemn at least one half of mankind to continued poverty, malnutrition and unemployment.' With reference to at least one . . . specific country in the less developed world:
> (a) State what you understand by the term Green Revolution.
> (b) Examine the causes of the problems of poverty, malnutrition and unemployment facing people in the less developed countries referred to above.

Teachers face a dilemma. There is a strong argument that such issues are of major global importance and therefore every geographer, even if immature, should examine them. But, if superficially treated, they lead to what one report stigmatised as:

> ... naive, over-simplistic accounts of the horrors of life and death in the squatter settlements around the main cities of Latin America and Asia. Few candidates ever seem to realise that at least some of the urban migrants not only make a living but actually prosper.

Within the CSE a national survey is hardly possible, for there is essentially a localised entry and exam system. Many exciting developments have taken place in schools where teams have developed their own courses and produced a Mode III assessment. Here surely is the true place for regional geography to survive – in the *local* surroundings at a variety of scales. However, of the 180 000 CSE entrants, the largest group is still for the Mode I, externally set, exam. There it does seem that an older regional model lingers on, with the risk of leading both teachers and pupils down a path that is comforting in its familiarity but a cul-de-sac as far as further study goes.

A levels have recently undergone waves of change, with the larger boards introducing reformed syllabuses within a few years of each other. These shifts do far more than dethrone the region. Quantification and theory, which is based on well-known models, both assume leading roles. Systems, albeit in simple form, are finding a place too. There is provision for individual field study and, in the formal exam, a swing away from the long prose essay towards shorter, more structured, questions. No longer need the candidate

> Consider whether regional contrasts within the British Isles are becoming more or less pronounced.

He must '(a) Suggest a specific urban land use for each of the 3 bid rent curves ...' in a given diagram and must, '(b) Justify your choice in each case.' All answers to be written within the space provided on the answer sheet. The freedom of 'Answer three questions' has tended to give way to restrictions such as 'Part 1 core studies (50 marks) will be examined by means of ONE compulsory section of short answer questions ($1\frac{1}{4}$ hours).' Teachers who value literacy and powers of expression deprecate these trends. Examiners would reply that their job is to discriminate fairly. The more precise the questions, and the greater the comparability of candidates' answers, the more equitable the result will be.

What can be discerned about the near future? Certainly A levels are becoming more scientific and exact:

> Describe and explain how *you* would collect data to test whether, in the area shown on the OS map provided, there is a relationship between slope angle and (a) geology, (b) aspect, (c) altitude...

But this has happened just at the time when social and behavioural studies are gaining ground strongly in universities. The costs and consultations involved in reform make any further major changes unlikely before the late 1980s. It seems that some young geographers may be climbing aboard a quantified or systematised craft just when others are saying that it is not the only one worth sailing in.

Although most GCE Boards practice extensive consultation with teachers, the A levels still lay great stress on preparation for university courses which only a small minority of the 36 000 candidates will ever enter. One of the exciting features of the Normal (N) and Further (F) syllabus designs was that commissioned groups could take a blank sheet and say 'Suppose we made a fresh start: how would we want the average 18-year-old to see our subject?' Sadly the prospective modular syllabus of the West Midlands Teachers Group and the attempted plan to combine spatial and regional theory with environmental awareness of the Oxford and Cambridge Board group have been laid on the shelf.

The main hope of immediate reform again rests with a Schools Council Project, this time for the 16 to 19 age group. Their framework for a Man/Environment syllabus was strongly backed by a recent questionnaire to a small sample of students and over 75 per cent favoured an approach that was relevant to daily life and experience with a compulsory practical element. More significantly, the project has drawn together enthusiastic groups of schools in very different parts of the country. Not until the mid-1980s though will those outside the project really begin to be able to make use of its materials and ideas.

One aspect of both this scheme and the Bristol Project will arouse controversy if widely adopted in the 1980s. This is the large part played by the internal assessment of course or fieldwork in the final grading. On the one hand is the view that teaching is a graduate profession and therefore its members have a right to play a big role in any assessment process. Coupled with this is the view that two or three brief spells under pressure may not do justice to candidates' real abilities. On the other hand is the view that any individual teacher has too limited an experience of standards to see beyond a local pattern or be consistent from year to year. Whereas it is felt that the exam boards have resources to attain a national or regional view and their experience allows them to maintain a stability which validates their results for employers and universities. There is no doubt that any widespread replacement of external exams by internal assessment would greatly increase teachers' loads and the resultant moderation needed would involve costly absence and travel.

The burden is manifest in other ways. There is a constant filtering down of concepts for the teacher to simplify and present. In 1973 the A level student was asked:

> How would central place theory help in planning the settlement structure of a newly developing area?

But by 1979, the O level aspirant was asked to:

> Discuss the part played by central place theory as an aid towards the understanding of how rural settlements evolve. What are the limitations of this approach?

My personal experience suggests that such requests are at, or beyond, the pupils' theoretical understanding and yet this conceptual load has to be grafted on to a subject which still carries heavy factual luggage.

In another board this was the syllabus for *half* one GCE O level paper in 1979:

> Section A: systematic human geography
> The following is suggested as an outline for this part of the course:
>
> The distribution of population; types of settlement; the chief functions of towns; the conditions of production of the major items of food and raw materials which enter world trade (studies of the more important examples on cereals, vegetable oils, fibres, animal products, sugar, rubber and timber); sources of power; manufacturing industries, with special reference to the British Isles; world communications.
> (Where appropriate, examples taken from Europe should be studied.)

Is there really time for this and for an approach to quantification so that the A level candidate can avoid complaints that:

> . . . few candidates recognised Chi Square as a suitable test and they misapplied it by using absolute values rather than frequencies for 'observed' and 'expected values'? (From an Examiners' Report.)

Although the GCE boards have all pruned their syllabuses in the course of reform one feels that yet more could go overboard.

The problem is exacerbated because elsewhere in schools geography as a discipline is being eroded. Integrated studies are prominent up to 13 years of age with the contributions of the discipline sometimes strangely muted, although the best results rival the funded projects in their vitality and success. Environmental studies appear in exam syllabuses, though the entries are, at present, small. In one CSE board geography attracts 18 000 per year, its new rival only 2 000.

The divorce between physical and human parts of geography is something long foreseen by critics. But the marriage has so far held. Whether Gould's vision of a split, with an essentially problem-solving human geography emerging will become as true in the UK as it is in the USA is dubious (Gould[1]). Gould saw geography rooted in powerful theoretical concepts and armed with sharp mathematical instruments. But are these not tools for the graduate rather than the school child?

Many other topics of study have been suggested for the next decade. Smith has presented a case for the introduction of welfare study:

> The question of who gets what *where*, and how provides a framework for the restructuring of human geography in more socially relevant terms without necessarily abandoning the vigour and sophistication of the quantitative era (Smith[2])

an argument which is made fully elsewhere in this book. Lynch looks forward to an age in which leisure and city life predominate (Lynch[3]). His vista opens up new roles in agencies structuring the passage of time, giving information about space and time, preserving the past and living 'in the middle range future'. Is all this fantasy? Some would dismiss it as such. Or is it a more realistic field of study for a Britain with over 2 000 000 unemployed than:

> Select one of the following: sugar cane, maize, cotton.
> (a) Name an area where production is important and draw a sketch map to show its location.
> (b) Describe the physical and other conditions which favour its production in that area?

Certainly if we do not allow ourselves to speculate freely about alternative futures for geography there is a danger that the subject, and its exams, will ossify into a quantitative or theoretical mould, just as surely as it did into a regional one twenty years ago. Looking at the pace of change in the subject and the rapidity with which pupils' lives alter, such fossilisation would clearly be disastrous. Continued evolution of the subject and its exams may displease the teacher who asked at a meeting that the 'new O level should not be altered for twenty years', but it should make a more stimulating life for most of us and our pupils.

Acknowledgements

My thanks are due to the following for permission to reprint quotations from their Examination Papers or Regulations:

The University of Cambridge Local Examinations Syndicate, The University of London University Entrance School Examinations Council, The Oxford and Cambridge Schools Examination Board, The Northern Ireland GCE Board, The Schools Council 14–18 Bristol Geography Project.

References

1 GOULD, P. R. (1973) 'The open geographic curriculum', in *Directions in Geography*, ed. R. J. Chorley. Methuen

2 LYNCH, K. (1972) *What Time Is This Place.* MIT Press
3 SMITH, D. M. (1974) 'Who gets what where, and how: a welfare focus for human geography,' *Geography* **59**(4)

Table 6 *Geography entries at A level, 1965–77*

	1965	*1968*	*1971*	*1974*	*1977*
Total A level entries	370 435	417 822	456 996	477 753	554 448
English	40 281	50 698	58 860	62 243	67 291
	10·83%	12·13%	12·88%	13·03%	12·14%
Physics	43 396	41 593	41 791	41 868	45 086
	11·71%	9·95%	9·14%	8·76%	8·13%
Pure and applied maths	26 051	31 526	36 082	36 526	47 790
	7·03%	7·55%	7·90%	7·65%	8·61%
History	30 580	34 377	35 992	36 374	39 166
	8·26%	8·23%	7·88%	7·61%	7·06%
Geography	24 436	29 179	32 817	34 137	35 933
	6·60%	6·98%	7·18%	7·15%	6·48%

N.B. About 1 in 10 A level candidates in geography goes on to specialise in the subject in some form of higher education.

Table 7 *Geography entries at O level, 1965–77*

	1965	*1968*	*1971*	*1974*	*1977*
Total O level entries	2 170 019	2 161 720	2 223 826	2 477 320	2 852 045
English language	348 688	341 134	334 945	398 871	472 622
	16·07%	15·78%	15·06%	16·10%	16·57%
Mathematics	234 289	226 210	230 080	247 594	289 665
	10·80%	10·46%	10·35%	9·99%	10·16%
English literature	199 173	202 774	212 547	234 469	253 976
	9·18%	9·38%	9·54%	9·46%	8·91%
Biology	128 285	140 898	157 136	187 032	221 183
	5·91%	6·52%	7·07%	7·55%	7·76%
Geography	160 765	156 613	159 474	178 587	196 960
	7·41%	7·24%	7·17%	7·21%	6·91%

N.B. Entrants for the 16+ examination are included in both GCE and CSE statistics

Table 8 *Geography entries at CSE level, 1965–77*

	1965	*1968*	*1971*	*1974*	*1977*
Total CSE level studies	230 977	836 480	1 180 892	2 179 779	2 782 297
English	41 487	149 092	200 253	400 561	526 880
	17·96%	17·82%	16·96%	18·38%	18·94%
Mathematics	38 802	132 494	186 100	327 314	413 749
	16·8%	15·84%	15·76%	15·02%	14·87%
History	16 743	58 665	79 150	132 772	156 846
	7·25%	7·01%	6·70%	6·09%	5·64%
Geography	21 177	69 364	93 951	153 553	181 028
	9·17%	8·29%	7·96%	7·04%	6·51%

N.B. 1 16+ examination statistics are included in both GCE and CSE tables
2 In 1977, the 16+ entry in geography was 15 781
3 In 1977, the CSE Geology entry was 6 136 (0.22%)

Table 9 *The 'top twenty' subjects in popularity in a regional CSE Board* (*East Anglia*), *1979 figures*

1	English	44 802	11	Chemistry	10 248
2	Mathematics	39 538	12	Woodwork	7 067
3	Art	19 758	13	Metalwork	7 025
4	History	19 925	14	Social studies	6 274
5	Geography	17 930	15	Technical drawing	5 674
6	Physics	15 255	16	Arithmetic	5 665
7	French	14 758	17	Office practice	4 860
8	Biology	14 451	18	Religious education	4 440
9	Typing	13 191	19	German	4 318
10	Home economics	10 326	20	Human biology	3 715

Other figures of interest; Environmental Studies 1 036, European Studies 963, Geology 494

Table 10 *The impact of the funded projects*

a The 'Bristol' 14–18 Project (O level examination administered by the Cambridge Board. Candidates can enter through any GCE board).

	Number of candidates	*Number of schools*	*% of UK O level geography entry*
1974	386	10	0.22
1975	624	12	0.34
1976	2 298	44	1.23
1977	4 604	91	2.44
1978	4 417	89	2.40
1979	5 604	115	3.00 approx.
1980	5 648	124	3.25 approx.
1981 (estimated)	5 700	154	3.30 approx.

N.B. 1 At least another 1 000 candidates take CSE examinations based on 'Bristol' work.

2 Full statistics for O level candidates beyond 1978 are not officially available at the time of compilation of this table (1981).

b The 'Avery Hill' Project (O level examination administered by the Welsh and Southern Boards. Candidates can enter through any GCE board. There are many associated CSE syllabuses).

	Number of O level candidates (estimated)	*Number of CSE candidates (estimated)*	*% of UK O level entry*
1977	275	500	0.18
1978	1 500	5 000	0.80
1979	3 800	10 000	2.08
1980	5 500	20 000	3.20
1981	8 500	44 000	4.65

Table 11 *A level entries from certain subjects in the Cambridge Local Examinations Syndicate 1976–80.*

	1976	*1977*	*1978*	*1979*	*1980*
Total entries	51 251	54 877	55 808	56 221	57 700
Geography	4 913	5 244	5 290	4 900	4 831
History	4 830	5 061	4 825	4 780	4 898
Economics	1 924	2 162	2 321	2 397	2 630
Geology	357	424	357	316	303

N.B. The table of total entries excludes private candidates and subjects with less than 100 candidates.

Table 12 *O level entries from certain subjects in the Cambridge Local Examinations Syndicate 1976–80*

	1976	*1977*	*1978*	*1979*	*1980*
Total entries	292 364	299 478	298 003	333 848	312 578
Geography	25 858	24 736	28 983*	32 999*	27 224**
History	22 131	23 220	22 559	22 654	22 364
Economics	3 505	3 695	3 767	4 286	4 916
Geology	–	–	–	–	–

N.B. The table of total entries includes private candidates but not all minority subjects.

* Includes candidates in the 16+ feasibility pilot study

** Excludes candidates in the 16+ feasibility pilot study

4 The overlap between school and university geography

Vincent Tidswell University of Hull

At a time when planned progression and continuity in the curriculum are rightly receiving so much attention, it may seem somewhat perverse to examine discontinuity.

But despite the recent assertion of Bailey[1] that 'Geography has come of age' there remain a number of disturbing discontinuities in our educational system which affect what is ultimately achieved. Three such discontinuities may be readily identified:

1 Between primary and secondary schools. What assumptions are made about the knowledge and skills acquired in a child's previous school? Perhaps only a brave teacher would make any such assumption in the light of HMI *Survey of Primary Education*[2].

2 Between O and A level. To what extent does the O-level examination with its heavy reliance upon factual recall really prepare students for the demands of A level? There can be no doubt about the rigour demanded at this higher level in terms of skills, methods of enquiry and conceptual understanding.

3 Between secondary and higher education. Problems of overlap at this interface have been debated in *Geography* for more than a quarter of a century. Usually the discussion appears as a report of the Sixth Form and University Committee of the Association (see *Geography* 1975), the most recent of which concerned itself with

essential skills and techniques (see *Geography* 1979). In the letter, not only are the demands of current A-level syllabuses analysed and commented upon, but also the Working Group sets out its own ideas deemed to be appropriate at this stage.

It is suggested that all three of these discontinuities impede a smooth progression in geographical education and each demands not only attention but urgent action if continuity and progression in the curriculum are to have any true meaning. However, attention here is focused upon the third of these 'unconformities'.

Throughout time, the university discipline has always influenced what has been taught in schools. Such influences may be traced from the regional tradition of Herbertson to current emphasis upon spatial analysis which owes so much to the work of Peter Haggett. This interrelationship between university and school is a natural and healthy one: how else could what we teach be revitalised by new ideas and new approaches?

Whilst change in the longer term is evolutionary in nature, the rate of change is rarely constant. From the middle 1960s we have experienced rapid change which has excited some but alarmed many. The impact for schools of changed philosophies and fresh approaches to familiar problems first manifested itself in the sixth form. A new and demanding syllabus was introduced by the Oxford and Cambridge Examination Board in the late 1960s. How quickly this pioneering work has been followed up by almost all A level Boards is evident from Bowler's[3] analysis of current requirements. Problems of translation of examination syllabuses into teaching ones have been tackled in a professional way as exemplified by Fyfe et al[4] and Silson[5]. Consumer reaction to what has been unfortunately and mistakenly dubbed 'The New Geography' is expressed by an A-level candidate, Catherine Smith[6].

Concern for A-level attainment levels is not confined to the public examination boards. Objectives for sixth form geography were identified by Slater and Spicer[7] as a contribution to the 16–19 Schools Council Project. Humphreys[8], in attempting to discover what qualities are looked for in university departments in their new students, found any specific or criteria-referenced answer elusive, largely because of the diversity and independence of universities.

We are alerted by both Perkins[9] and Unwin[10] to the credibility gap which may well exist between the aspirations of A-level syllabus statements and the actual attainment levels of candidates. Unwin in particular challenges the wisdom of implanting 'sharp tools' (some quantitative methods) into 'unskilled hands' and cites as evidence the competence of his third year undergraduates, some of whom will of course be tomorrow's teachers. He argues cogently for a reasonable study of 'exciting and relevant' theory rather than knowing 'how to estimate a beta coefficient in order to draw a line through a scatter of points'. This view of Unwin echoes the ideas of Kirby and Lambert[11] who also plead for a postponement of quantitative methods.

Unwin's challenge on attainment levels is courageous and timely, but is based on a small and skewed sample. It was interesting to read in the same number of *Teaching Geography* a necessarily anonymous letter exposing levels of ignorance from A-level answers between 1976 and 1979. Two examples may be amusing but they are also terrifying:

> C. Taller has developed a model town in his Central Place Theory. This idea of Chris Taller is ...
>
> The hydrological cycle describes three major activities of a river in its course

What then are the attributes of incoming undergraduates who have attained a satisfactory grade at A level to read for an honours degree in geography? What foundations may be reasonably built upon and what major gaps exist? In an endeavour to establish some hard valid evidence rather than rely upon random anecdote, with the kind co-operation of colleagues, I collected data from 375 students in nine British universities.

Evidence was sought via a questionnaire, carefully structured to cross check respondents' answers, to ascertain reading habits, acquisition of statistical skills and competence to apply them, but above all to evaluate conceptual understanding of models and their underlying processes. Carried out in the first week of the first term of the 1978 session, this survey is stage one in the evaluation of recent change. As may be expected, the composition of the population is skewed towards the more able as defined by A level grades: 51 per cent having Grade A; 24 per cent Grade B; and 13 per cent Grade C.

Reading habits of the group are perhaps predictable. Easily the most frequently read journals are *The Geographical Magazine* and *The National Geographical Magazine*, with *Geography* ranking a poor third. A few people look at *The Geographical Journal*. Whilst this pattern may merely reflect the availability of journals in schools, it is tempting to draw the inference that the sixth former remains interested in people and places rather than rings and hexagons.

Clearly for the universities a major task is to teach in-coming students to read, interpret and appraise research articles upon which will depend the effectiveness of much of their future work.

A second area to be investigated was that of statistical skills and techniques. Each student was asked to recognise specified items and show awareness of their usefulness, application and limitations. The results are displayed in Table 13 and show that recognition of log graph paper and how it transforms data were both appreciated by a majority. The most fundamental data-collecting device, namely sampling, reveals an impor-

Table 13 *Acquisition and application of skills*

Skill competence	*% of all respondents*
Recognition of log graph paper	69
Appreciation of usefulness of log graph paper	60
Recognition of sampling grid	43
Appreciation of significance of standard error	8
Application of nearest neighbour analysis	19
Application of chi squared test	24
Interpretation a regression line	36
Full appreciation of regression line	24

tant weakness, with a mere 8 per cent understanding of the method. Only 19 per cent were familiar with nearest neighbour analysis as a means of seeking a measure in a dot distribution and less than 25 per cent knew when to apply the chi-squared test in a given appropriate context. Establishment of relationships between phenomena is a significant part of much geographical enquiry. Commonly expressed as a regression or 'best fit' line, such lines appear in many contemporary texts for the main school, yet only 36 per cent could reasonably interpret a regression line and a mere 24 per cent could identify problems the line did not solve.

Perhaps it is fair to conclude therefore that whilst many students seem to have acquired a basic facility in recognising and using a range of statisitical techniques, few know how to interpret the results of their labours or appreciate the limitations or opportunities of the methods used. The findings of this enquiry tend to support David Unwin's 'hunch' based on his one set of observations.

Given the inclusion of simpler models in A level syllabuses, it is reasonable to expect the first year undergraduate to have a working knowledge and understanding of them. Models used in the enquiry included those of von Thunen, Christaller, Burgess and Hoyt, being deemed the most common ones.

Naming the usual diagrams of von Thunen and Christaller presented no difficulty to 81 per cent of the candidates, but there was a marked fall off when understanding the workings of the model were probed. Less than 50 per cent could determine the k value in the diagram presented, whilst less than 20 per cent knew what the k value indicated. In fairness there was a marked difference between A grade candidates and others. The two questions posed above were answered correctly by 65 per cent and 26 per cent respectively by A grade students compared with 28 per cent and 5 per cent by all other grades. When models were combined as in the case of Hoyt and Burgess in Fig. 22, correct identification dropped to 48 per cent. Indeed, many students made wild guesses, writing down all the urban models which sprang to mind.

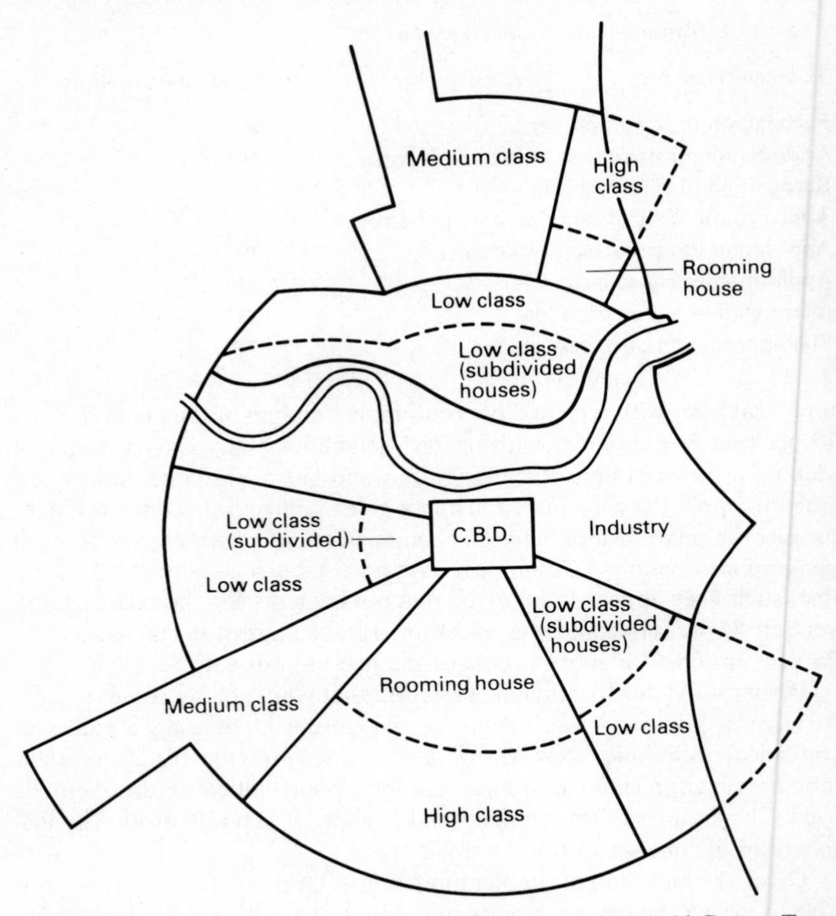

Fig. 22 *Idealised model of ecological areas in Sunderland* (Harold Carter, *Towns of Wales*, University of Wales Press, 1966)

The ability to identify the concepts which underpin models was disappointing, yet it is precisely this ability which needs to be encouraged both from intellectual and educational standpoints. Overall 52 per cent understood the basic idea underlying von Thunen's model, 26 per cent that of Christaller whilst 28 per cent showed an appreciation of the influences at work in urban models. One very popular topic is that of retail geography. When confronted by Wrexham's zones of influence (Fig. 23) some 20 per cent could suggest the use of Reilly or Huff models to generate a generalised sphere of influence but fewer than 10 per cent could identify in map form the spatial expression of the range of a good. Allied to this problem of compartmentalised thinking was a failure to recognise that fundamental ideas such as distance decay can be applied in more than one context. Summary results are recorded in Table 14.

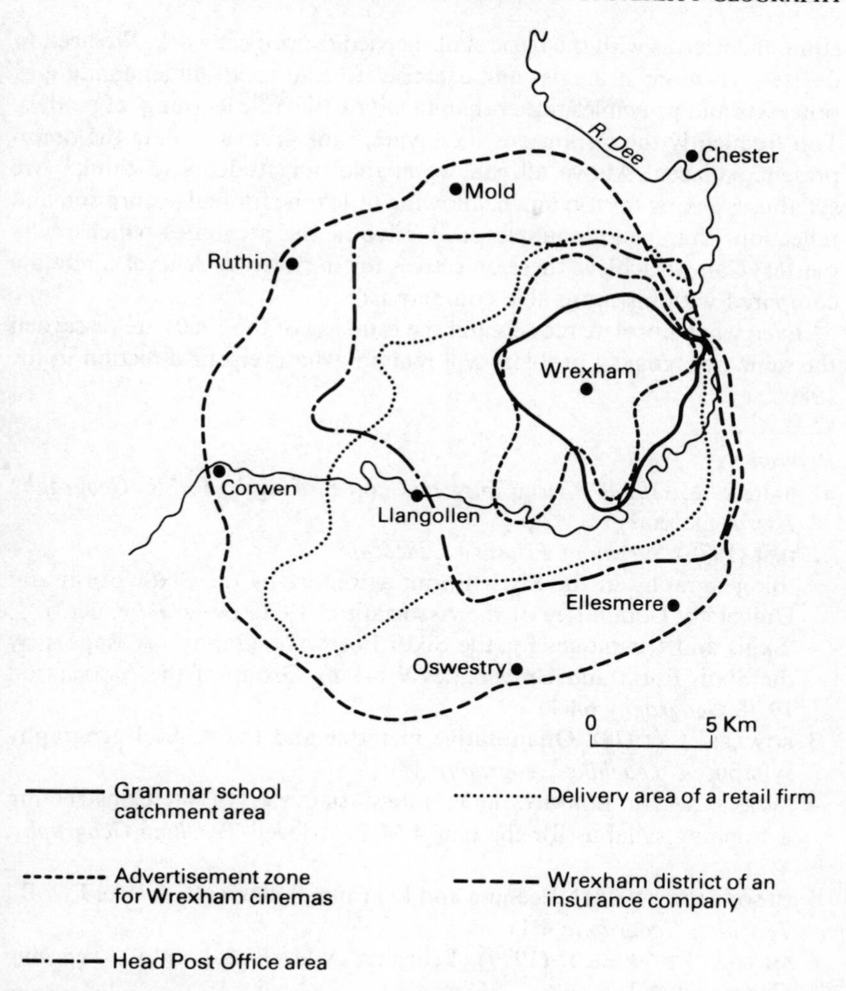

Fig. 23 *Wrexham: Zones of influence* (Harold Carter, *Towns of Wales*, University of Wales Press, 1966)

Although the results of this enquiry may at first seem disappointing to curriculum innovators, encouraging progress has been made compared with a decade ago. However, there is no room for complacency. In particular there is a need to stimulate wider reading and to ensure students

Table 14 *Models and concepts*

	Grade A	*Others*	*Overall mean*
Ability to recognise models	90%	73%	81%
Understanding of underlying concepts	48%	29%	40%
Application of concepts in wider context	38%	11%	25%

equip themselves with the basic skills needed for project work. We need to devise even more strategies and exercises to lead to an understanding of processes and principles rather than facilitate the 'role learning' of models. Too frequently the response is Pavlovian: rank size rule being the omnipresent panacea. Above all can we enable our students to think? Are syllabuses asking for too much, allowing little time for real absorption and reflection, thus being counter productive in the attributes which really matter? Can we achieve these objectives for the average A-level candidate compared with his more able counterpart?

Even when obsolete models and the thinking of the 1960s are discarded the same pedagogical problem will remain, whatever our direction in the 1980s.

References

1 BAILEY, P. (1979) 'Geography has come of age', *T.E.S. Geography Extra*, December

2 HMI (1978) *Survey of Primary Education*
'Biogeography in the Sixth Form; a Report by the Sixth Form and University Committee of the Association' 1976. *Geography*, **60**(3)
'Skills and techniques for the Sixth Form Geography'. A Report by the Sixth Form and Universities Working Group of the Association 1979. *Geography* **64**(1)

3 BOWLER, I. (1978) 'Quantitative methods and the A level geography syllabuses' *Teaching Geography* **3**(3)

4 FYFE, ELSPETH; HORNBY, BILL; JONES, MELVYN (1977) 'Constructing a teaching syllabus for the new J.M.B. A-level' *Teaching Geography*, **4**(2)

5 SILSON, A. L. (1978) 'Teaching and Learning Syllabus B, A level J.M.B.' *Teaching Geography* **4**(1)

6 SMITH, CATHERINE J. (1979) 'Learning A-level geography', *Teaching Geography* **4**(3)

7 SLATER, FRANCES and SPICER, BRIAN (1977) 'Objectives in Sixth Form Geography towards a consensus', *Classroom Geography*, December

8 HUMPHREYS, GRAHAM (1977) 'What university geographers expect of their new students'. *Teaching Geography* **3**(2)

9 PERKINS, PETER J. (1980) 'Building on sand', *Teaching Geography*, **5**(4)

10 UNWIN, DAVID (1980) 'Statistical inferences?', *Teaching Geography*, **15**(3)

11 KIRBY, A. and LAMBERT, D. (1979) 'Geography at school and university: is the gap between them growing?', *Papers on Geography in Education* (2), Department of Geography, University of Reading.

5 Seven reasons to be cheerful? ... *or school geography in youth, maturity and old age*

Andrew Kirby University of Reading; **David Lambert** Ward Freman School, Buntingford, Herts

The only geography

In a recent article, Patrick Bailey (the Editor of *Teaching Geography*) identifies a new maturity within school geography: 'at last, after more than a decade of hard thinking and experiment, the shape of the new geography in schools is clear, as are the appropriate methods of teaching it. All we have to do now is to make best practice general' (Bailey, p. 31[1]).

There are, he asserts, seven reasons upon which to base the assessment. These are summarised below.

Sphere of activity	*Examples*
1 Establishment of theoretical foundations	The distillation of ideas into easily available books (unnamed)
2 Reassessment of methodology in geographical teaching	Examples: Bailey, *Teaching Geography;* Graves, *Geography in Education;* Hall, *Geography and the Geography Teacher;* Marsden, *Evaluating the Geography Curriculum*
3 Resources and materials for school use	Main school: *Oxford Geography Project* *Location and Links* Sixth form: *Patterns and Process in Human Geography* *Process and Pattern in Physical Geography* *Human Geography: theories and their applications*
4 Schools Council curriculum projects	Avery Hill Project Bristol 16–18 Project
5 Teachers' journals	*Teaching Geography*
6 Reorganisation of A level	All examining boards
7 Morale	'The feeling abroad that geographers now know where they are going'

There are several points of detail with which we would initially take issue. Potentially the most serious is the glossing-over of the central problem of the subject's theoretical foundations (paragraph two of Bailey's discussion). The precise nature of these foundations is not outlined; the suggestion appears to be that a consensus now exists upon their definition and that their diffusion is now complete. However, without some statement as to the nature of these accepted principles, it is impossible to gauge the accuracy of the claim. This is compounded by the fact that no sources are given from which guidance may be obtained.

A second point relates to the geographical journals for teachers. Although Bailey observes that *Teaching Geography* has brought the Geographical Association back in touch with its members, it would be over-optimistic to regard the relationship as smooth. In a recent editorial, Bailey himself writes that 'it is difficult to get these kinds of "teaching" articles, yet readers continually ask for them' (Bailey, p. 98[2]): this suggests that many teachers are still in the role of adopters, rather than innovators vis-à-vis the changing syllabus. (A subsidiary point with which we would take issue here is Bailey's re-writing of history. Most geographers would pay some tribute to Neil Sealey, who single-handedly established and developed *Classroom Geographer*. It was the latter which showed that, in Bailey's words, the GA was out of touch with its membership, and that an alternative to *Geography* was required.)

Our third point relates back to the question of theoretical foundations. In his final paragraph Bailey concludes that 'the "new" school geography has come of age ... indeed it is the *only* geography' (Bailey,[1]; original emphasis). His use of terms such as maturity, and the general tone of his argument cause us to infer that Bailey regards a transitional period to have come to an end; more important, he does not admit to the likelihood, nor even the need, for any imminent changes to the recently-completed edifice of school geography. This is in our view insubstantiable. If we accept Bailey's analogy concerning stages of development, it is clear that maturity has replaced a period of youth and vigour, and this will in turn be replaced ... by old age and protracted senility. It is clearly unrealistic to suggest that what exists at a particular time can be 'the *only* geography', and that change is undesirable. This is unarguable on intellectual grounds alone. There are, however, new tensions arising within the subject as a whole, which also necessitate continued change. We will discuss these in turn.

Geography in school

In this section, we aim to consider contemporary geography in the sixth form; in the main school, other issues are of interest, but our general remarks here will of course also have implications for lower down the school. Essentially, our observations fall into two categories.

The first concerns the expansion of the sixth form, or what is sometimes referred to as 'the new sixth'. The ability range in the sixth form is wider than before, which has implications for the choice of materials and teaching strategies. The changing nature of the audience demands that quotations such as that used a decade ago to herald the first Charney Manor conference still require serious consideration:

> most of the stuff that I have read about the 'new geography' has been either well above school level or else so full of waffle that how people have applied it in school is beyond my comprehension.
>
> (WALFORD, p. 1[3])

The problem facing middle-ability students (who after all constitute nine-tenths of the sixth form) is essentially one of a perceived lack of relevance in much of the syllabus. Many take up geography in a spirit of interest about the world, yet increasingly much of the material with which they are presented does not relate to reality as they know it. Even as far as most texts are concerned, there is a basic consensus that, for example, urban morphology can only be understood in the context of American cities and settlement patterns in relation to southern Germany.

Here we are identifying a basic problem in the way in which information is transmitted throughout a discipline. The 'new geography' has taken very selectively from the different pigeonholes that make up the subject as a whole. The criterion for selection appears to have been as much as anything else the ease of extraction and presentation. This has manifested itself as an emphasis upon pattern rather than process, Chicago rather than Human Ecology, hexagons rather than Central Place Theory. For the lower-ability candidate this represents a massive exercise in redundancy, whilst for the able student a very garbled view of the world is the likely outcome; Michael Bradford for example has demonstrated that only small proportions of those who take A level to a standard that gets them into university have anything approaching a 'correct' understanding of very basic geographical concepts (Bradford[4]).

Our second concern relates to the basic intellectual challenge offered by an A level course at present. One of the most telling criticisms of geography pre-1970 was that it lacked not only rigour, but more importantly intellectual stimulation. No one could deny that rigorous analysis is now *de rigueur*; whether the introduction of, for example, quantitative methods has been accompanied by an increase in such stimulation is open to question. As an illustration of this we reproduce in full a question from the 1979 Cambridge A level practical paper:

'The following tables refer to cereal production in parts of the United Kingdom in 1974. (Source: Agricultural Censuses and Production 1974.

Table 1: *Areas under tillage (crops and fallow) (Units: Millions of acres)*

	Wheat	*Barley*	*Oats*	*Other*	*Total*
England	2.9	4.3	0.4	2.5	10.1
Scotland	0.1	0.9	0.2	0.2	1.4

Table 2: *Average wheat yield in selected countries (Units: cwt. per acre)*

County	*Yield*	*County*	*Yield*
Avon	35.7	Nottinghamshire	41.2
Bedfordshire	43.0	Oxfordshire	36.2
Cheshire	34.8	Shropshire	37.2
Derbyshire	43.0	Staffordshire	40.2
Gloucestershire	37.9	Warwickshire	37.3
Leicestershire	37.9	Worcestershire and Herefordshire	37.6

Average of the tabulated values: 38.6

(a) From the data in Table 1, prepare two divided circle diagrams (pie-graphs) showing the composition by crops of the areas under tillage in England and Scotland respectively. So that they may be compared more readily, both diagrams must be drawn on the same page.

(b) (i) From the data in Table 2, determine the standard deviation of the values from their average.

(ii) Discuss briefly the merits of the standard deviation as a supplement to the average value.'

An example such as this can be criticised in three ways. First, it is not analytical; the candidate is not asked to apply himself or herself to a substantive problem. Second, the question is in no sense geographical; the data could have been generated by random numbers. Third, the question reduces quantitative matters to trivial concerns, that can be answered with stock phrases learnt by rote. In this sense, it probably contains less pedagogic value than an analysis of that old favourite, a synoptic chart, which is also to be found on the same paper.

To summarise, therefore, we argue that all within the garden is not quite as rosy as Patrick Bailey would suggest, that contemporary geography is not the only geography, and that evolution is necessary. This is the case even without a consideration of changes within the university sector, to which we address ourselves below.

Geography at university

Another substantial problem facing school geography is the growing gap that exists between it and the research frontier of the subject, a topic that we have considered at some length elsewhere (Kirby and Lambert,[5] Fisher[6]). Theoretically, there is no necessity for practitioners of school geography to take an interest in the antics of some researchers, who seem more concerned to determine whether or not Kant and the Venerable Bede were geographers than with substantive issues. This ignores the power structure of the discipline on the one hand, in which examining boards are controlled by university academics, and the intrinsic importance of developments within higher education on the other; as we shall argue here, the present deliberations within the subject have a clear message for school practitioners.

An overview of the subject at present is very difficult, as the discipline is in a state of flux. Nonetheless, three clear pointers can be identified.

The first is a decline in the relative importance of quantitative methods. This can be measured in various ways, but a useful pointer is the fact that the Institute of British Geographers Quantitative Methods Study Group is no longer the largest such in the IBG (that mantle has fallen to the Urban Study Group). Researchers entering the discipline do not now concentrate upon methodological issues to the exclusion of subject matter: but more important, even amongst those still active in quantitative work, there is a growing uncertainty about future developments. A large minority, for example, is increasingly keen to move back from high-pressure quantification towards the far simpler approach offered by exploratory data analysis (this is discussed in Kirby[7]).

In a recent paper, John Bale has gone further, and has questioned the extent to which we are on the brink of an 'anti-quantitative putsch'; he writes, 'the time is ripe to apply some of Taylor's conditions for overthrowing an academic discipline to radicalism, just as he had applied them to quantification' (Bale p. 2,[8] Taylor,[9]). He concludes that, 'it goes without saying that the potential implications of the radical revolution for the geography profession are far greater than those of the quantitative revolution a decade or so before' (Bale p. 48).

There are two components to this argument. The first is that geographers have become greatly involved in what we might call substantive issues, be they in relation to the identification of social and economic problems, or the formulation of policies and remedies. Again, a simple indicator of this would be a study of some of the IBG Transactions Theme Issues: on *Housing and People in the City*, on *The Inner City*, and *Geography and Public Policy*, for example.

The second is that geographers have realigned themselves within the hierarchy of disciplines. Researchers no longer look automatically to physics as a source of methodological and epistemological ideas (Wilson[10]); instead they are turning to sociologists and philosophers, as Derek Gregory argues in his book *Ideology, Science and Human Geography*. His work has ended the decade by placing a major challenge before the discipline in terms of how it should conduct its research in the future (Gregory[11]).

To summarise these remarks, therefore, we may argue that the research section is undergoing change, and that this process has major implications for school geography. The chairman of the IBG Quantitative Methods Study Group has recently written that, 'as one of the quantifiers of the mid-1960s and as one who has spent much of the past 13 years teaching and promoting their use in tertiary education, can I raise some serious doubts I have about this continued incorporation into secondary education?' (Unwin, p. 144[12]). In other words, half the discipline has moved on, and is now looking over its shoulder at the other half – with, it must be said, some disquiet.

Conclusions

In this paper we have argued that Bailey's optimistic remarks concerning the maturity of geography are unfounded. Our case may be summarised by two dovetailed quotations:

> Since school teachers form the largest component of the geographic profession it is not only vital that they are aware that radical geography is emerging, but also that it displays many of the features associated with the changes Taylor described for the sixties.
>
> (BALE, p. 4[8])
>
> It is the subject's saving grace that its current research, literature and teaching now reveal a continuing state of change which far from being disruptive is the necessary progress previously absent.
> The significance of these developments is surely that the future of geography is progressive rather than established.
>
> (SEALEY, p. 2[13]).

References

1 BAILEY, P. (1979) 'Geography come of age', *Times Educational Supplement*, 7 December, p. 31.
2 BAILEY, P. (1980) 'Editorial: an article rejected', *Teaching Geography* **5**(3), 98
3 WALFORD, R. (ed.) (1973) *New Directions in Geography Teaching*, Longman. London
4 BRADFORD, M. (1977) 'Some problems of varying overlap between

secondary and higher education', *Journal of Geography in Higher Education* **1**(1), 80–6
5 KIRBY, A. M. and LAMBERT, D. M. (1978) 'Geography at school and university – is the gap between them growing?', *Papers on Education in Geography* **2,** University of Reading
6 FISHER, G. C. (1979) 'The relationship between school and university courses in physical geography', *Papers in Geography* **3,** Bedford College, University of London
7 KIRBY, A. M. (1979) 'The state of the science: quantitative education in British Universities and Polytechnics', *Papers on Education in Geography* **3,** University of Reading
8 BALE, J. (1978) 'How to overturn a discipline again', *Classroom Geographer*, December, pp. 2–5
9 TAYLOR, P. J. (1976) 'An interpretation of the quantification debate in British geography' *Transactions* IBG, NS **1**(2), 129–42
10 WILSON, A. G. (1972) 'Theoretical geography; some speculations', *Transactions* IBG, **57,** 31–45
11 GREGORY, D. (1978) *Ideology, Science and Human Geography*, Hutchinson. London
12 UNWIN, D. (1980) 'Statistical inferences', *Teaching Geography* **5**(3), 144
13 SEALEY, N. (1978) 'Editorial' *Classroom Geographer*, November, p. 2

6 Geography and the school curriculum debate

Richard Daugherty University of Swansea

For most of the 1960s and 1970s discussions among geography teachers were mainly concerned with the *what*, *how* and – to a lesser extent – *why* of geography teaching. It is only since the mid-1970s that the debate has broadened and the question of *whether* geography should be taught to all pupils in schools has come to the fore. Increased attention to the school curriculum by politicians, educational administrators and the general public has challenged teachers to explain and justify their contribution to education. What are the implications of this debate for the teaching of geography? How have geography teachers responded to the challenge? How might changing circumstances in the increasingly political arena of the school curriculum affect the subject in the 1980s?

The late 1970s marked the end of a period in England and Wales when ministers and officials of central government disclaimed responsibility for the school curriculum. From the 1940s, when Tomlinson remarked that 'Minister knows nowt about curriculum', through to the end of the 1960s, when Crosland accepted that neither he nor his officials were 'in the slightest degree competent to interfere with the curriculum', the ministers responsible for the education service concentrated on organising and financing it. It is true that Eccles regretted the preoccupation with the structure rather than the content of education and expressed concern about the 'secret garden of the curriculum'. But Eccles's 'commando-type unit', the Curriculum Study Group, was short-lived and soon superseded by a teacher-controlled Schools Council for Curriculum and Examinations. During the post-war period local education authorities (LEAs), though legally responsible for the curriculum, normally delegated that responsibility to school governors who, in turn, usually allowed headteachers a free hand in this as in most other aspects of internal school organisation.

By the late 1970s politicians, HM Inspectorate and Department of Education and Science (DES) officials were all playing their part in ensuring that discussion of the school curriculum was no longer seen as the private concern of headteachers and, perhaps, their colleagues. Prime Minister Callaghan's Ruskin College speech in October 1976 is normally regarded as marking the re-emergence of interest in the curriculum among politicians. Subsequent changes in government and in Secretaries of State for Education have not brought any diminution of that interest. Prompted by the politicians, DES officials drafted consultative documents such as *Educating our Children. Four subjects for debate* (DES[1]) and *Education in Schools. A Consultative Document* (DES[2]), in both of which the school curriculum loomed large. Perhaps it is not surprising, given the lack of attention to the curriculum by officials over a long period, that when proposals for consultation emerged in January 1980 as *A Framework for the School Curriculum* (DES[3]), the failure of officials to appreciate the complexity of the issues involved was all too apparent in the many rough edges of that document.

HM Inspectorate had also taken on a more public role over the same period with published surveys of primary and secondary education in England (DES [4,5]) and 'working papers' on *Curriculum 11–16* (DES[6]) which were offered as 'a contribution to the current debate'. The Secretaries of State subsequently invited HMI to 'formulate a view of a possible curriculum' which, as *A View of the Curriculum* (DES [7]), was published at the same time as the *Framework* proposals.

Ministers and officials have been at pains to point out that legal responsibility for the curriculum remains with the LEAs, while also arguing that 'there is a need to review the way these responsibilities are exercised' (DES p. 1[3]). Thus the purpose of DES circular 14/77 and the resulting report, *Local Authority Arrangements for the School Curriculum* (DES [8]), was innocently described as 'to collect information from local education authorities about their policies and practices in curriculum matters' (p. iii).

Teacher protests that there were dangers in the invasion by politicians of an area of professional expertise were met by the argument that politicians had a duty to ensure that the education service met national needs. As Lawton put it: 'The state has a duty of spelling out the supposed advantages of schooling and this will involve explicit statements about curriculum content' (Lawton, p. 26[9]).

The school curriculum thus became a matter for national debate as well as being argued out, often in terms of timetabling rather than of educational philosophy, by teachers within each school. Teachers of geography no less than teachers of other subjects continue to press at school level for what they regard as adequate provision for the subject. But the national debate will inevitably influence curriculum decisions in schools both indirectly and, more obviously, where reference to the 'importance' of an aspect of curricular provision in a national document is quoted as justification for a controversial timetabling decision.

How have geography teachers responded to the call for justification of the subject's place in the curriculum? The statements of aims which were formulated as part of the burgeoning of subject-based curriculum development in the late 1960s and early 1970s were seen by some cynics as mere decorations, the wrapping to be attached to the curriculum package once geography teachers had decided amongst themselves and according to their own criteria what that package should contain. However, as pressures grew on teachers to account for what they were doing, such statements, whether general in character (Geographical Association[10]; HMI[11]) or attached to courses in individual schools (Howarth[12]), were recognised as the public face of the subject, the point at which the non-geographer could begin to interpret and assess the nature and value of the subject's contribution to education.

The demand for justification came at a time of fundamental rethinking of the subject in schools. *New Directions in Geography Teaching* (Walford (ed.)[13]) pointed the way, followed by several books articulating and discussing the way forward, with the most significant feature of the change in emphasis being summed up in the title, *The Teaching of Ideas in Geography* (DES[14]). But, with the decisions on course objectives and content remaining in the hands of individual teachers, the pace of real change in school geography was very gradual. Although the influence of national curriculum projects cannot be measured only in terms of the numbers of candidates for examinations related to them, nearly a decade after the launching of two successful Schools Council projects for the 14–16 age group fewer than 45,000 candidates sat project-related 16+ examinations in 1979 out of a total of 360,000 candidates for 16+ geography examinations. Thus those responsible for statements justifying geography's place in the curriculum not only faced the perennial problem of the dated perception of the subject which curriculum decision-makers (actual and aspiring) have, but also had to explain a subject very much in a transitional phase.

Perhaps it is typical of a geographer to think first in terms of where the curriculum debate is located – as national as well as at school level – but it would be unwise to ignore the associated shift in the terms and tone of the debate. Growing concern about the country's economic performance led some to lay part of the blame on the schools and to look to them for 'solutions'. Also, pressure on public service finances encouraged a 'value for money' outlook – 'getting better value for the resources which can be afforded' (DES p. iii[8]). The pressure of greater accountability came in various forms and at all levels, including for example demands for more parent governors and for greater 'choice' of school. Disillusionment with an era of expansion and teacher-led innovation was widespread, finding voice in the Black Papers and the popular press. The view that 'the curriculum is too important to be left to teachers' gained currency – and that quotation came not from one of the Black Papers but from a Fabian Society pamphlet entitled *Whose Schools?* (Corbett[15]).

Graves reports that in France in the atmosphere of crisis after the Franco-Prussian war part of the blame was attributed to inadequate geography teaching: 'our general staff had made gross geographical errors and misread their maps' (Graves, p. 49[16]). In the atmosphere of economic crisis in Britain in the 1970s geography teaching, far from being held partly to blame, was scarcely mentioned in the succession of official documents leading up to the *Framework* proposals. In the HMI publications, geography was not one of the subjects included in the *Curriculum 11–16* working papers. Few readers of that document will have noted the comments that the selection of subject interests discussed was 'neither exhaustive nor exclusive' (DES, p. 19[6]) and seen it as an adequate counter to the implied significance of the twelve subject areas which were discussed. Similarly in *A View of the Curriculum* the assertion that there was 'a strong case for maintaining some study of history in the final secondary years' (DES, p. 18[7]) was not matched by a consideration of the case for extended study of geography. Indeed, more stress was laid on the value of geographical studies in the chapter on the primary school curriculum in spite of the well-charted decline in geography teaching at that level in recent years (DES[17], DES[4]). The HMI proviso that too much should not be read into what are intended to be purely illustrative selections and comments may not be heeded by the hard-pressed headteacher searching for any higher authority to guide him through the curriculum battlefield within the school.

DES publications likewise gave little prominence to geography in discussing the elements of the curriculum. Certain subject areas were identified as having 'recently been topics of general concern' (DES, p. 5[8]). The fact that geography was not among those topics might have brought some comfort to teachers of the subject if much the same list had not become the basis for a 'core curriculum' outlined in the *Framework* proposals. There the list of 'core' subjects was justified either because they 'gave rise to special problems' or because of their 'intrinsic importance for all pupils'. By a strange coincidence the subject areas which had given rise to recent discussion and concern were also those which were 'intrinsically important'. Geography was listed with the 'additions to the core subjects' on the last page of the proposals.

Why should geography, part of the existing core of subjects taught to all pupils in the early years of the secondary school and a major option for 14- to 16-year-olds, be almost invisible when politicians, administrators and HMI took the initiative in opening out the curriculum debate? Daugherty and Walford[18] (1980) suggested several possible explanations. Could it be that geography teachers had paid insufficient attention to explaining their aims, except to each other? Or could their explanations have been noted by politicians, parents and pupils but found wanting? Perhaps learning about the world, deemed important in the days of Empire and still for many the main reason for teaching geography in school, is no longer thought to be a major function of schooling in an era of more insular attitudes and of television as a powerful alternative source of the individual's world picture? There is no doubt some truth in all these arguments. Until the mid-1970s there were few attempts to explain school geography to a wider audience and those who tried to do so only partially succeeded in finding the right language in which to convince the outsider who was interested in the outcome of education and not much impressed by the claims of the educationist. More recently an editorial in *Teaching Geography* (Bailey[19]) restated familiar claims: geography is *the* environmental study with 'unique integrating potential', 'a distinctive education in thinking', 'an education for the future'. But were those claims accepted, or even heard, by the participants in the wider curriculum debate?

However, to look only to the quality and frequency of statements about school geography for an explanation of the apparent lack of public recognition of its value is surely too myopic an outlook. It ignores the fact that many other significant areas of the existing school curriculum were no more prominent in the debate than was geography – music, history and craft for example. Disillusionment with educational innovation has led to an atmosphere of retrenchment, of 'back to basics'. The hazy notion of a 'core curriculum' is hardly questioned; the debate turns on what should be included in the 'core', on emphasising the 'basics'. Even Denis Lawton[20], a leading advocate of curriculum reform, is reduced to nostalgic reflection on the qualities of the 1937 *Handbook of Suggestions for Teachers*, though he does baulk at the handbook's neglect of social, political and economic studies. In an atmosphere of retrenchment writers such as Maurice Holt[21] can be dismissed as idealists when they point to the inadequacies of a secondary curriculum dominated by the grammar school tradition and argue the need for a curriculum designed specifically for the comprehensive school. 'The stridencies of backlash educational politics raise the so-called basics to insistent consciousness, glibly equating the teaching of science with the harnessing of technology and environment and the teaching of modern languages with greater exporting power' (Daugherty and Walford[18]).

It is not difficult to pick holes in the case being made for a national curriculum 'framework' as the Geographical Association response to the DES proposals showed (Geographical Association[22]). Of course 'a reasoned justification' is needed of the choice of curriculum elements allocated 'core' status and 'guidelines on curriculum structure should include a statement of aims and agreed cross-curriculum concerns'. The fact remains that the proposals were welcomed by a wide range of interest groups from the universities to the TUC. The crude assertions and muddled thinking of the proposals were less important to many than the mood which they represented – a feeling, unimpaired by lack of evidence, that schools had neglected the 'basics' and followed every educational fashion, however wild and ephemeral. The main thrust of the debate was not concerned with what was interesting or valuable in education but with what, in a vague and ill-defined way, was indispensable. One of the few hopeful points for geography teachers in the DES proposals was the reference to 'helping pupils to understand the world in which they live, and the interdependence of individuals, groups and nations' in a list of six possible aims for schooling (DES[3]). But when it was put to DES officials that the proposed 'core' seemed to neglect that aim, the reply came that the 'core' which they had outlined was not related to those particular aims nor indeed to any statement of aims, the production of which would be beyond the scope of DES responsibility. It seems that, when the mood is right, a firm basis for proposals can be dispensed with; unsupported assertion is in, logical argument out.

Publication of *The School Curriculum* (DES[23]) marked the end of a phase of central government initiatives on the curriculum. Whereas the *Framework* proposals were 'preliminary views', *The School Curriculum* was offered as 'guidance to the local education authorities and schools in England and Wales on how the school curriculum can be further improved'. It maintains the position that not only are there curriculum questions which all schools should ask themselves but there are 'right answers' to be found to such questions. The possibility that there could be several different, but equally valid, answers does not appear to have been considered; the accusing finger is pointed at 'unacceptable diversity' and standardisation of provision is put forward as the cure-all. The strong case for a common curriculum for all the pupils in a particular school is thus confused with the more dubious case for a uniform curriculum in all schools. Furthermore, although the assertions in *The School Curriculum* are less baldly and less boldly stated than those in the *Framework* proposals, its conclusions rely more on unargued preferences than on reasoned propositions. For example. 'The sections which follow do not cover exhaustively every subject or aspect of the curriculum but seek rather to focus on certain elements which the Secretaries of State wish to emphasise at the present time.'

To many people in schools, preoccupied with a discouraging economic environment, the DES guidance represents no more than a passing irrelevance; yet another document pronouncing blandly on what should be, while the schools struggle to maintain what is, in the face of diminishing resources, inadequate staffing and with many school leavers destined for unemployment. Any geography teacher with the time to read *The School Curriculum* might take some comfort from the place accorded to the subject in the primary school and the recognition of its part in the 'broadly similar programmes' (the term 'core' having apparently gone out of (fashion) of the first three years of secondary schooling. Some of the specific recommendations also deserve attention. For example, are all lower secondary geography courses designed so that pupils who study the subject no further 'are enabled to achieve something of value' (para 41)? However, if *The School Curriculum* is remembered at all it will not be for such details but for its part in defining the terms of the national curriculum debate. Although the governemnt is no nearer to finding an agreed basis for determining a uniform, common curriculum, the pressure is maintained for convergence towards a government-led consensus.

What then are the options and prospects for geography in the school curriculum in the 1980s? In so far as the ultimate responsibility for curricular decisions is likely to remain in the hands of the staff of each school, much will still depend on what is actually achieved with pupils. With falling numbers on the rolls of many schools, however, there will be few opportunities to remedy weaknesses such as those reported in the HMI primary by additional staff or other resources. There will also be cases where non-replacement of staff hits subjects such as geography, deemed by some to be 'non-essential', harder than most.

One option of course would be, at a time when the utilitarian view of the curriculum prevails, to play the utilitarian cards. A geography course could be devised, aimed at, for example, helping children find the local Job Centre or, should they happen to be sailing in the South Atlantic, making sure they knew when to turn left for India. An alternative and more productive approach in an era when 'value for money' is the popular cry, would be to work for acceptance of a more liberal interpretation of 'value' or 'worthwhileness' in relation to education. If geography can help pupils 'gain understanding of issues and problems as well as of places and peoples' (Geographical Association[22]), must it not have a place in the education of all young people?

But to argue only in those terms is to follow the familiar path from the subject to the curriculum as a whole. What the opening out of the curriculum debate has invited is a discussion which starts with the whole curriculum and then moves on to the details. For the moment, apart from the 'areas of experience' mooted in the HMI Curriculum 11–16, the discussion of details has been in terms of conventional subjects, reflecting both the conservative character of the debate and the lack of accepted alternative terms in which to discuss the components of a curriculum. But, once the question of the whole curriculum is on the public agenda and not just the preserve of curriculum theorists, the field is open for alternative ways of answering that question. As Becher and Maclure[23] have argued, we may have passed through a period of subject-based curriculum development and be moving into a phase of system-based curriculum development.

At least three possible consequences of the changing circumstances and climate of curriculum thinking can be envisaged for the teaching of geography. An emphasis on the curriculum as a whole could highlight the weaknesses and gaps in existing provision and lead to the introduction of additional curriculum components and the extension of teaching in some subject areas. An inevitable consequence would be the reduction in time allocated to established subjects such as geography. Acceptance of the case for geography in the education of all young people does not in itself justify the subject being taught to all for the first three years of secondary schooling and to some 40 per cent of the age group for a further two years.

A further possibility is that combined studies courses may be given fresh impetus as a way of 'filling the gaps' in the subject curriculum, of providing 'more coherent learning experiences' for pupils and of making 'more flexible use of staff'. In an era of system-based development and of some dissatisfaction with a traditional subject pattern, combined studies courses could prove an increasingly attractive option for curriculum decision-makers. Viewed from the perspective of the system, whether at school or national level, the subject teacher's objections to the replacement of separate subject teaching by a combined studies approach can be portrayed as nit-picking and concerned with self-preservation. Geography teachers will need to take account of the case for such courses and to be positive in their contribution to them rather than being preoccupied with the supposed threat to the subject's and their own identity.

Whether or not combined studies courses increase in number, a system-based perspective on the curriculum can also be expected in another respect to hasten the demise of the view of the curriculum as little more than an amalgam of subject contributions. A shift in perspective to the whole curriculum serving the 'needs' of the pupil and of society, combined with pressure for accountability, may focus the argument about curriculum content with geography more on what the subject is *for* and less on what it *is*. The 1960s and 1970s saw the geography teacher in school hanging on, with difficulty, to the coat-tails of the academic geographer and looking mainly to the academic subject for guidance on how the geography course in schools might change. Sterile arguments about whether to teach rank correlation in the second year of the secondary school obscured the real issues of the contribution of the subject to the general education of all young people. If the opening out of the curriculum debate places the focus firmly on the aims of a general education for all (no doubt interpreted in very different ways) and puts subjects in their proper place as vehicles for, rather than the principal sources of, those aims, school geography will not necessarily be the poorer for it.

References

1 DES (1977a) *Educating our Children. Four subjects for debate.* HMSO
2 DES (1977b) *Education in Schools. A Consultative Document*, Cmnd. 6869. HMSO
3 DES (1980a) *A Framework for the School Curriculum.* HMSO
4 DES (1978b) *Primary Education in England. A survey by HM Inspectors of Schools.* HMSO
5 DES (1979b) *Aspects of Secondary Education in England. A survey by HM Inspectors of Schools.* HMSO
6 DES (1977c) *Curriculum 11–16: Working papers by HM Inspectorate.* HMSO
7 DES (1980b) *A View of the Curriculum. HMI series: Matters for Discussion 11.* HMSO
8 DES (1979a) *Local Authority Arrangements for the School Curriculum: Report on the Circular 14/77 review.* HMSO
9 LAWTON, D. (1980b) *The Politics of the School Curriculum.* Routledge & Kegan Paul
10 GEOGRAPHICAL ASSOCIATION (1977) 'Geography in the school curriculum', *Teaching Geography*, **3**(2), 50
11 HMI (1978) 'Geography in the School Curriculum', *Teaching Geography* **4**(2), 76
12 HOWARTH, B. (1977) 'Opting for geography in the fourth year', *Teaching Geography* **3**(2), 67
13 WALFORD, R. (ed.) (1973) *New Directions in Geography Teaching.* Longman
14 DES (1978a) *The Teaching of Ideas in Geography. HMI series: Matters for Discussion 5.* HMSO
15 CORBETT, A. (1976) *Whose Schools?* Fabian Research Series 328
16 GRAVES, N. J. (1975) *Geography in Education.* Heinemann
17 DES (1974) *Education Survey No. 19. School Geography in the Changing Curriculum.* HMSO

18 DAUGHERTY, R. and WALFORD, R. (1980) 'Overlooked in the Secret Garden', *Times Educational Supplement*, 18.4.80
19 BAILEY, P. (1979) 'The geographer's contribution', *Teaching Geography* **5**, No. 2, 50
20 LAWTON, D. (1980a) 'Common core lessons from the class of "37"', *Times Educational Supplement*, 4.1.80
21 HOLT, M. (1978) *The Common Curriculum*. Routledge & Kegan Paul
22 GEOGRAPHICAL ASSOCIATION (1980) 'Government proposals for the school curriculum', *Geography* **65**(3), 232
23 DES (1981) *The School Curriculum*. HMSO
24 BECHER, T. and MACLURE, S. (1978) *The Politics of Curriculum Change*. Hutchinson

Part Three

Signposts for the future

Introduction

The contributions of this final section do not represent sequential argument. They can be read in any order. They represent a variety of signposts into the future and there is little doubt that the arms of the signposts point in several different directions. None of them points in the direction from which we have come, however. One road is visible from another along the way; others take different routes to a generally similar horizon.

Derek Gregory provides a significant viewpoint from the academic frontier; his powerfully argued paper will repay considerable re-reading since it compresses much argument and implication into the confines of a short contribution. It lays the foundations for an important new perspective in geographic thought and draws on a view of thought in social theory which is likely to be one which classroom teachers will encounter more frequently in subsequent years.

Gregory does not reject the gains made through the sharp and clinical positive analyses of space in the 1960s and 1970s; but he sees them as possessing only a limited adequacy. They have provided some cohesion to geographical thought; but they are not enough to fully explain the deep structure of the real world. 'It is not enough to have the processes of social science operate on the real world – they must operate *in* it'.

His final paragraphs are a thoughtful challenge to those who would seek to duck low from the firing-line after ten years hard evolutionary labour, and also to those who would seek to cut school geography adrift from its academic counterpart. 'It may be, of course that it is all too difficult ... but this would be a curious response from a profession which has complained for so long that it is not taken seriously enough.'

John Huckle shares a similar disenchantment with the strong positivism of the past decade, but concentrates his attention on some of the psychological, sociological and pedagogical aspects of such a view. He argues for more attention to be given to educating pupils in the affective domain. It is not enough merely to express a *concern* about values, says Huckle; geographers were respectfully touching their forelock in that direction when the predecessor of this book was compiled. What is needed is a commitment to certain value positions and a teaching strategy to implement this. Like Gregory, Huckle is influenced by writers and thinkers in the realms of social theory who are relatively unknown to classroom geographers as yet.

The concern of both Huckle and Gregory is in harmony with the points made by Richard Daugherty in his paper in Part Two. If geography is to continue to have a place in the school curriculum of the future, it will have to furnish external credibility as well as internal consistency; is the way forward to return to a 'lighter', informational type of study, or seek to use an understanding of theory to probe ever more deeply for the *'real'* explanations of spatial and environmental phenomena?

Trevor Bennetts tills a quite different soil. He builds on the deeper understanding of conceptual linkages which the last decade has brought to link progression in geography to notions of pupil development. Progression is a term to which many pay lip-service, but it has always lacked many credible examples; Bennetts here provides a detailed and comprehensive analysis and example of how it can successfully work in practice. It makes some presumptions about the continuing place of geography as a study throughout the secondary school. . . .

David Walker and Michael Day turn the spotlight on methodology in the next decade. Walker is well placed, as the administrator of the Geographical Association's Package Exchange (GAPE), to identify the potential of the computer, and of the micro-processor. Day's balancing piece look at the problems of using educational technology in the classroom and colourfully brings the problems of day-to-day planning and management into view.

Eleanor Rawling asks that geographers play a greater part in developing links with others who, like them, are interested in the environment. This is not merely a plea for a shift in content, but a reminder that environmental education 'is concerned equally with skills, and with feelings and emotions, and it must involve direct experience'.

The implications, therefore, for methodology are significant. In several of the chapters in this section, as in the lesson units in Part One, there is a continuous desire to develop pupil participation in an active way and to take seriously the viewpoints which the learners hold.

In the midst of movements towards more flexible and adventurous forms of assessment, Gerry Hones sounds a cautionary note. Those who would assess by coursework and informal methods must be prepared to be thoroughly expert and professional in the way in which they tackle the job; and it is a considerable extra aspect of the teacher's role to do this.

Rex Walford's chapter is an attempt to clarify some of the viewpoints of contemporary discussion, by identifying certain recurring ideological standpoints which lie at the heart of styles of geography teaching.

The first part of the chapter seeks to de-mythologise some of the 'official language' which clouds such positions by its blandness, or by its careless use.

Walford suggests that if geography teachers take some steps to consider their own ideological positions, and those of others, discussion about the future of the discipline may be conducted more honestly, more pungently and more profitably in the next decades.

7 Towards a human geography

Derek Gregory University of Cambridge

'Outside the university entrance ... I slowed and began glancing in at the porticoes and pillars, the formal pompousness of the place, and I was thinking that such impersonality, formality, is how one can most easily identify a place of learning - school, university, college, and that this atmosphere must set a condition of thinking for a young person being educated in it. I saw a man come down those steps, but this was a time for people to be going home, and there was a steady stream of them coming across to the gates. I was looking at them idly and thinking how tinily unimportant these human beings looked beside the great cold buildings that were supposed to be their servants, and that no young thing learning there could ever believe that human beings are more important than their institutions ...'

DORIS LESSING: *Briefing for a descent into Hell*

Recapitulation

The 'new geography' has, with an almost vicious rapidity, become the 'old geography'. It is still possible to find those who, like the handful of Japanese soldiers who emerged from the jungle in the 1970s, still fighting the long-lost battles of the Second World War, refuse to acknowledge the new alignments and continue to campaign for the victory of their old allegiances: but these are now a minority. Most geographers have recognised that the quantitative revolution of the 1950s and 1960s was not the first time their subject had changed, and neither would it be the last. But this is not to suggest that the history of the subject is one of a succession of 'paradigms', because this would inevitably conceal the *continuities* between one phase and the next and the *contradictions* within each of them, and it would also suppress the *connections* between changes in geography on the one hand and changes in its constitutive society on the other. I don't want to provide a critique of the Kuhnian model, which is readily

Note:
Derek Gregory had intended to be present at Charney 1980 but was prevented from attending the conference in person at the last moment. A first draft of his paper was thus discussed in his absence. Some weeks later, he met a group of Charney participants and there was further discussion of the paper. This revised draft is the outcome of that evening's deliberations.

available elsewhere,[1] but I do think it important to spell out these three elements in more detail in order to provide some sort of context for the argument which follows. In doing so, I will have to take a number of short-cuts and deploy a series of short-hand expressions, and so it must be remembered that my descriptions here can be little more than crude caricatures. Even so, the outlines should be firm enough.

In the first place, then, it is essential to realise that the rejection of the positivism which characterised much of the geography in the 1970s need not entail the rejection of quantitative methods as such and that, on the contrary, an intellectually viable geography cannot afford to negate the genuine advances made during the quantitative revolution. Now it is certainly true that the central assumptions of positivism can be connected up to the requirements of particular analytical procedures: for example, positivism is predicated on a naive empiricism in which observation statements are assumed to be independent of theoretical constructs, and if this is challenged – as it is by most modern philosophies of science – then the whole edifice of conventional hypothesis testing will have to be radically restructured.[2] But this need not undermine the foundations of statistical inference itself. And in fact the last ten years have witnessed the emergence of mathematical and statistical methods of extraordinary power and subtlety, particularly through the impressive contributions of R. J. Bennett, A. D. Cliff, P. Haggett, J. K. Ord and A. G. Wilson, which have enabled geographers to interrogate space-time data-sets of considerable complexity in a much more rigorous and incisive fashion.[3] These continuities should, I suggest, be strenuously maintained and integrated much more closely into the developing corpus of geographical inquiry.

At the same time, however, they are not a substitute for it, and it is no longer the case that the concerns and concepts of the subject can be articulated in these narrowly technical terms. Indeed, and in the second place, there is an increasing tension between this complexity of *method* and the triviality of the pre-existing *models*. Harvey spoke directly to this when he identified 'a clear disparity between the sophisticated theoretical and methodological framework we are using and our ability to say anything really meaningful about events as they now unfold around us'.[4] Most of the classical models were directed towards the elucidation of equilibrium spatial structures, and their translation into geography involved a reinstatement of its 'geometric' traditions: as Bunge put it in his *Theoretical Geography*, 'the science of space finds the logic of space a sharp tool'.[5] Yet in many ways it was much too sharp: it sliced right through the specificities of social life, and cleaved away the pulsating rhythms of social reproduction and historical transformation, to expose an invariant, abstract geometry – a lifeless skeleton, shorn of its human flesh. This sort of intellectual surgery always had its critics, of course, and the anti-vivisection lobby has grown in recent years; but it now derives much of its impetus not from the stout defence of the Hartshornian orthodoxy, but rather from the revival of interest in a social theory capable of addressing questions of comparable complexity.[6] It is still far too early to talk of an emerging 'consensus',[7] I think, and in any case geography may well never stop 'branching towards anarchy',[8] but a cluster of core concerns is nevertheless discernible, and these owe much to a critical engagement with modern Marxism. Geography was probably the last of the social sciences to take Marxism seriously, at least in Britain and North America, but even so it is now beginning to distance itself from its earlier, largely mechanical rehearsal of Marx's original compositions,[9] to leave behind its coterminous and unyielding recital of objections to them,[10] and to attempt a careful and considered examination of their contending claims. I suspect that nothing could be more wounding than a continued dialogue of the deaf, and that progress towards a genuine integration of human geography with social theory will necessarily depend on the sustained *negotiation* of the barriers between these discourses.[11]

Such efforts clearly demand more than an elemental 'putting the "human" back into "geography"', although the reinstatement of the humanistic tradition, with its affirmation of human creativity, sensitivity and agency, is not in any doubt; but at the same time it is also necessary to explicate the relations between the constructs through which this tradition might be reappropriated *and the wider society which gives them their effectivity*. This is both a past and a prospective exercise. We now know, for example, about the connection between von Thünen's land-use models and the crisis of European society in the early nineteenth century; we can identify the ways in which Weber's industrial location theory was tied to the cultural consequences of the industrial revolution in early twentieth-century Germany; and we can recognise the various forms in which the 'urban question' of inter-war America surfaced through the work of the Chicago School in general and of Robert Park in particular.[12] None of these projects was a 'free-floating' intellectual construction, therefore, and each of them was reciprocally related to an historically-specific social context. The reappraisals of these classical models are thus rather more than indications of a resurgence of interest in the history of modern geography – which is in itself remarkable enough[13] – since they are also moments of a deeper movement which is committed to a conjoint critique of contemporary materials. The purpose of this admission of *historicity* is to ground the theoretical discourses which are articulated through geography in their constitutive and historically-specific social structures: in short, to connect geography to the society of which it is a part and to which, in some sense, it remains responsible. This allows for the simultaneous interrogation of both sides of the equation, subject and society, and hence ought (in principle) to provide for a disclosure of the structural interests which lie buried beneath the by now familiar rhetoric about 'values in geography'.[14]

It should now be clear, I hope, that the whole thrust of this argument has been to prepare for some awkward questions about how the geography we teach represents (and is constrained to represent) the relations between human agency and social structure, and about how these representations reach back in to society to restrict (and legitimise the restriction of) the efficacy of human agency.

These are, of course, at once political and ethical issues and, as David Ley has demonstrated, while any geography which draws back from them is scarcely *non*-political it is profoundly *un*ethical: disengagement is not disinterest, therefore, and any attempt to suppress these issues is also an attempt (however well intentioned) to promote a definite set of (unacknowledged) political and ethical commitments. As David Harvey put it, 'any claim to be ideology-free is of necessity an *ideological* claim', and no geography teacher – whatever his conception of 'science' – can escape from this ideological imperative.[15]

Reconstruction

In the past human geography typically represented the relations between the individual and society in one of two ways. In the first of these modes, which is sometimes called 'reification', society was treated as a reality *sui generis* which is somehow external to and constraining on human agency, so that primacy was accorded to system and to structure. These sort of notions can, with little difficulty, be traced back to the Vidalian concept of the *genre de vie* (which clearly owed much to an encounter with Durkheim's collectivist sociology), but they were revivified in modern geography through both systems theory – where 'the whole' is (literally) 'greater than the sum of its parts' – and structuralism – where a series of structural Marxisms drawn from the theorems of Althusser and Piaget provided highly abstract concepts of the mode of production.[16] By contrast the second mode, 'voluntarism', treated society as being somehow constituted by intentional action and so promoted agency over system or structure. These counter-claims received their fullest expression in Weberian sociology, but they were readily translated into modern geography through an explicitly idealist programme and its subsequent extension into the language of phenomenology. The key-words in this new humanism, no matter what its grammar entails, radiate from a fundamental concern with the creative and affective dimensions of human agency: with the existential concepts of meaning, intentionality and subjectivity.[17] These twin oppositions are summarised in Fig. 24, which also indicates one way in which modern geography is beginning to resolve them through a theory of *structuration*.[18]

This formulation flows from Marx's claim that if men make society,

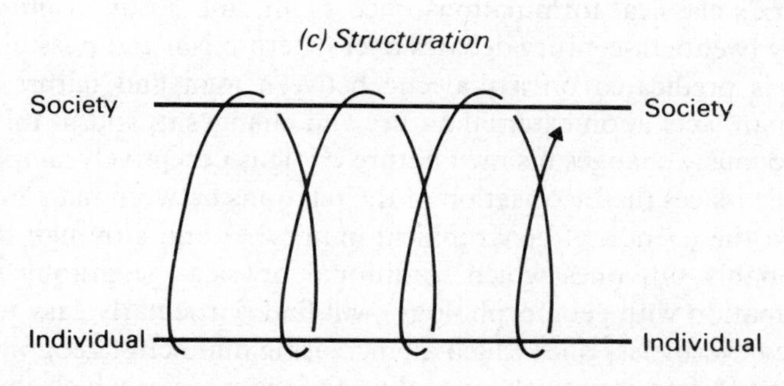

Fig. 24 *Reification, voluntarism and structuration*

they do not do so merely under conditions of their own choosing. Society is essentially and intrinsically historical, therefore, and contrary to the voluntarist position its structures pre-exist and post-date any individual constructions. But at the same time these social structures are not the constraints and barriers which reification supposes, but instead the very *conditions* and *consequences* of human actions. An example might make this clear: in speaking, I necessarily draw upon a pre-existing linguistic structure, and although I might not be able to specify the rules and resources which it makes available with any great precision its existence is nevertheless a *condition* of every intelligible speech act; and, symmetrically, these utterances necessarily reach back to reconstitute that structure, whose reproduction thus becomes an unintended *consequence* of every speech act. In short, there is an inescapable 'recursiveness' to social life through which action and structure are dialectically inter-related.[19]

This scheme is still being developed, but even in its present programmatic form it provides rather more than the 'transition between action and structure' which Curry has called for in his development of a theoretical geography,[20] and even more than an elucidation of 'the relation of society and individual' which Olsson sees at the bottom of 'all political and social scientific theories,[21] because it also allows for the simultaneous re-integration and reconciliation of geography's by now largely disassociated 'vertical' and 'horizontal' dimensions: its 'ecological' and 'locational' schools.[22] I should perhaps underline that this rapprochement is *not* important because of any guarantees it might provide for the survival of the subject; on the contrary, these two dimensions are vital for the integrity of any social science worthy of the name and cannot provide for the disciplinary autonomy to which Eliot Hurst has (I think correctly) objected.[23] I now want to consider these two axes in turn.

The ecological school: society and nature

A central theorem of structuration is society's relation to nature. Recalling Marx's classical formulations once again, and cutting right through the early twentieth-century debates over determinism and possibilism, the scheme is predicated on a dialectic between man and nature through which man 'acts upon external nature and changes it, and in this way he simultaneously changes his own nature'.[24] This deceptively simple phrasing at last places the theorisation of the relations between man and nature 'back on the agenda of geographical inquiry',[25] and although the terms are probably not ones which traditional physical geography (with its preoccupation with geomorphology) will find particularly easy to accept, they are nevertheless ones which an increasing number of geographers are beginning to find intuitively appealing for the ways in which they represent the physical environment as both *condition* and *consequence* of human action. Elements of this are to be found in the classical conceptions of *la géographie humaine*, of course, and it is hardly news to the landscape school of historical geography, which still commands widespread professional and popular support; but what is novel is that these representations are located at the very centre of an explicitly *social* theory and as such seem to offer a new understanding of 'the role and relations of physical geography' through the non-defensive development of a 'vital and relevant physical geography'.[26] But while the theory of structuration demands an explication of the environmental systems on which social life depends and on elucidation of the ways in which they are restructured through human actions, the continuous transformation of nature is evidently more than a matter of progressive (or regressive) environmental change: it is also and at the same time an *economic* process (organised through the instrumental systems of social labour) and a *cultural* process (organised through the communicative systems of social signification).

And it is *here* that man 'changes his own nature'. By this I mean that any society depends on ecological exploitation, so its economy necessarily revolves around resource appraisals and appropriations. These can, of course, take a number of different forms in different societies, and even within the same society different sections of the economy can be tied to the physical environment in different ways; but what they all have in common is that their environmental relations are embedded in the economy *and* incorporated into culture, where they are elaborated into a complex field of signs, symbols and images which together help to make as well as to mirror the reproduction of practical life. And here too individuals draw upon resources and rules made available by these economic and cultural structures, and so reconstitute them in a constantly flowing series of economic and cultural engagements.

I take this to be of considerable importance, because modern human geography has been characterised by the gradual divorce of economic from cultural geography - a division which is frequently all the deeper for being one of philosophy and method as well as one of substance[27] - and any scheme that provides for the treatment of economy and culture within a single and non-reductive system of concepts is to be welcomed. As Olwig has argued in a perceptive essay, there is no need to regard man as either 'a part of nature' or 'apart from nature' when it is perfectly possible to think in terms of a dialectic between the ever-extending appropriation of nature through social labour and its simultaneous incorporation into an ever-unfolding system of social signification.[28]

The locational school: society and spatial structure

A second, co-equal theorem of structuration concerns the space-time relations inherent in the constitution of social life, and here too spatial structure can be represented as both *condition* and *consequence* of human action. The principle is easy enough to understand, and in fact Soja has already described the 'social relations of production [as] both space-forming and space-contingent', but in practice most attempts to provide a system of concepts capable of maintaining a balance between spatial structure as both product *and* support of social life have loaded the scales more heavily on one side or the other.[29]

Spatial structure is most clearly represented as a *condition* of social life in Hägerstrand's 'time-geography', and the original formulations of structuration drew upon the early contributions of his Lund school to chart the ways in which space and time can be seen as *resources* on which individuals have to draw in order to complete particular projects. Their access to these resources - and hence their ability to realise their particular intentions - is determined in part through a grid of capability, coupling and authority constraints and in part by the competitive allocation of projects between the available pathways in space-time (Fig. 25).[30] In principle, therefore, the repetition of characteristic space-time paths, the 'freezing' of a spatial structure as individuals are routinely directed along particular pathways, can be connected back to the reproduction of an equivalent social structure and in particular to the ways in which spatial asymmetries like class, age and sex confer what Hägerstrand calls differential 'power over space and time'.[31] In practice, however, the move back from the individual to society and the cycle from consequence back to

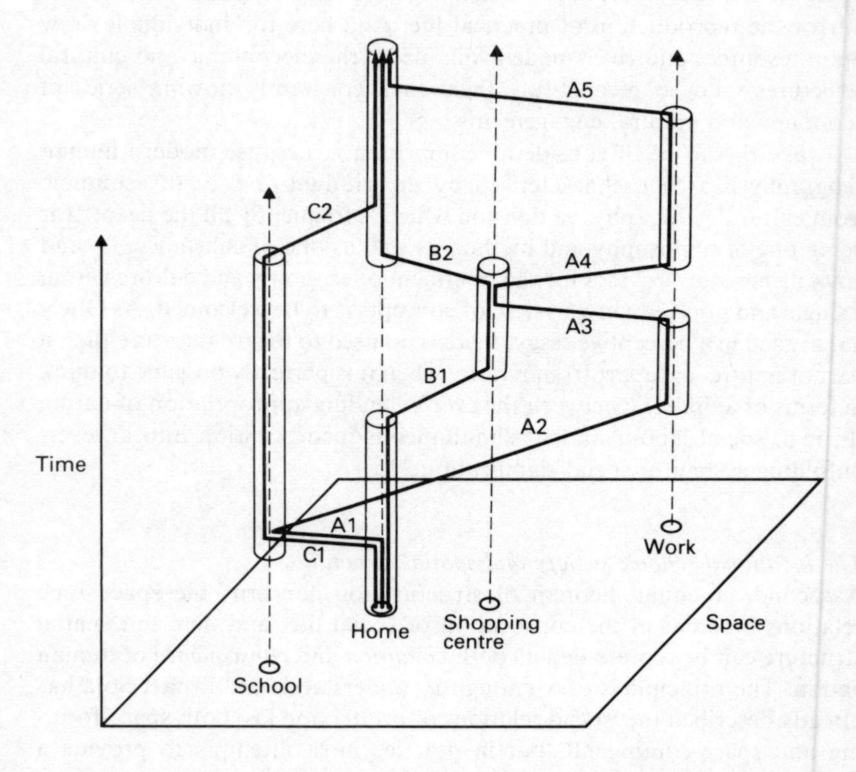

Fig. 25 *Space–time paths. A1/C1 husband takes son/daughter to school and then travels on to work (A2); wife travels independently to shopping centre (B1) and husband joins her for lunch (A3); he returns to work (A4), she to home (B2), where she is joined by her son/daughter (C2) and then her husband (A5). This example clearly makes a host of assumptions about the ways in which class, age and sex structure space–time paths, and in particular about their connections to capability constraints (e.g. use of private car by husband, public transport by wife), coupling constraints (e.g. lunch appointment) and authority constraints (e.g. school)*

condition remains strategically incomplete in most current versions of time-geography.[32]

Spatial structure is most clearly represented as a *consequence* of social structure in studies of 'combined and uneven development'. These attempt to specify the relations between the political economy of advanced capitalism and (in particular) the production and reproduction of regional and sub-regional inequalities. Such an approach entails a much deeper understanding of the internal workings of modern capitalism than conventional economic geography has been prepared to endorse, but whether one follows (say) Holland in his 'attempt to demonstrate that the trend of regional inequality is intrinsic to capitalist development'[33] or whether one prefers (say) Clark's more nuanced view that the connections between the two are contingent rather than necessary,[34] the analysis of spatial structure in these terms demands a prior explication of concepts which are almost entirely absent from the traditional vocabulary of economic geography. In many ways, I think, their absence undermined the intellectual credibility of these earlier analyses which, as I suggested earlier, never enabled us to say much more than the obvious (or the plain wrong) about the contemporary world, and it is a sign of the subject's maturity that it can now conduct serious discussions about the role of the state in the regional economy,[35] the geography of public finance,[36] the regional structure of inflation[37] and the impact of recession on the regional economy.[38] The list is only indicative, of course, and it can easily be extended: the ecological models of the Chicago School have long since been replaced by detailed discussions of the political economy of housing and of the relations between class structure and residential differentiation;[39] the classical models of diffusion processes have been overtaken by dissections of the ways in which the social structure (and in particular access to the means of production) can shape patterns of innovation adoption much more decisively than the geometry of information adoption.[40] But the point should be clear. And again, at least in principle, the social production of these spatial structures can be connected back to their recursive effects on social reproduction. Harvey, for example, has described 'the geographical landscape which fixed and immobile capital comprises' as 'both a crowning glory of past capital development and a prison which inhibits the further progress of accumulation', and this sort of characterisation can be applied at a number of different scales of analysis.[41]

It must be obvious, even allowing for my abbreviated account of them, that neither of these two schemes is complete and that both of them are still being developed; but for all their differences and difficulties - and these are considerable - moving beneath both of them is a considered reinstatement of *areal differentiation* as a central problem for modern geography. In the first edition of his *Locational Analysis in Human Geography*, Haggett complained that, largely as a result of the Hartshornian tradition, 'areal differentiation dominated geography at the expense of areal integration',[42] and his brilliant contributions have done much to redress (even to reverse) the balance. But if the wheel has now turned again, as I have been arguing here, it has not come full circle but rather moved in a spiral of which Bruner would be proud. The new focus on regional specificity - in Walker's splendid phrase, on the 'vast and intricate mosaic of unevenness'[43] - does *not* presage a return to a traditional regional geography. While these new developments retain many of the old commitments to places and the people that live in them, they also involve the deployment of clearly defined *theoretical* concepts to illuminate these particular regional conjunctures. In a sense, spatial analysis has become (in the most general and yet in the most profound terms) *social* analysis, and is at last beginning to provide for a truly critical understanding of the space-economy.

Reaffirmation

It may be, of course, that all this is 'too difficult': it clearly places enormous demands on the teacher. But this would be a curious response from a profession which has complained for so long that it is not taken seriously enough. And in any event the justification for a reconstruction of a subject ought not to be that it somehow makes it more respectable in the eyes of one's colleagues: the appearance of geographers capable of calling on computer-assisted techniques to solve particular problems in the subject may have dealt most satisfying body-blows to the sneerers who can be found in most common-rooms, but it certainly did not guarantee that the questions which they were asking were any more meaningful or that the answers to them were any more incisive. Indeed, a profession that worries so much about what its colleagues in neighbouring disciplines think is over-valuing the importance of the institutions through which it is taught, and it is small wonder that an obsessive concern with technical procedures should have left so little space for (and even *de*valued) the co-equal importance of human agency.

I have already made it clear that the advances of the Quantitative Revolution cannot be gainsaid; but neither can we stand still. The last ten years have seen a philosophical explosion, and while its shock-waves undoubtedly (sometimes perhaps unjustifiably) undermined the foundations of many pre-existing theoretical and empirical constructions, they nevertheless opened the way for a radical reappraisal of the subject: an interrogation of those questions and answers, and a clarification of its own 'conditions' and 'consequences'. These are tied in to society in a host of different ways, clearly, but there are good reasons to suppose that one of the effects of the Quantitative Revolution, particularly in schools, was a reduction of the subject to a narrowly technical education: to a set of mechanical exercises which were little more than occupational therapy and which provided for a labour force prepared for a routinised, repetitive labour process. The *models* rarely worked, and they certainly couldn't offer much of a critical understanding of the space-economy, but the *methods* were much more impressive. And at the same time 'science' was redefined so that it became synonymous with this technical conception: the ways in which scientific status was socially ascribed were rarely recognised, and the privileges accorded to it only occasionally challenged.[44]

But this sort of *critique* is, I believe, essential, for only then will we be able to create what elsewhere I called 'a doubly human geography: human in the sense that it recognises that its concepts are specifically human constructions, rooted in specific social formations and capable of – demanding of – continual examination and criticism; and human in the sense that it restores human beings to their own worlds and enables them to take part in the collective transformation of their own human geographies'.[45] These worlds are *not* isolated states or isotropic planes; they never were – and we have a duty to see that they never will be.

Note:

I owe a particular debt to Rex Walford, who over a period of several years now has been a never-ending source of enthusiasm and encouragement, and who has prompted me to think much more carefully about the educational issues involved in modern geography.

References

1 Discussions in the philosophy of science have been much more extensive (and much more critical) than most geographers have recognised, but see Johnston, R. J., *Geography and Geographers: Anglo-American human geography since 1945* (Edward Arnold, 1979), and Stoddart, D. R., 'The paradigm concept and the history of geography', in Stoddart, D. R. (ed.), *Geography, Science and Social Concern* (Basil Blackwell, 1981)

2 M. HESSE, *The Structure of Scientific Inference* (Macmillan 1974) and *idem, Revolutions and Reconstructions in the Philosophy of Science* (Harvester, 1980)

3 See, for example, Bennett, R. J., *Spatial Time Series* (Pion, 1979); Cliff, A. D., Ord, J. K., *Spatial Processes* (Pion, 1981); Haggett, P., Cliff, A. D., Frey, A., *Locational Analysis in Human Geography* (Edward Arnold, 1977); Martin, R. L., Thrift, N. J., Bennett, R. J., (eds), *Towards the Dynamic Analysis of Spatial Systems* (Pion, 1978); Wilson, A. G., *Urban and Regional Models in Geography and Planning* (Wiley, 1974); Wilson, A. G., *Kirkby*, M. J., *Mathematics for Geographers and Planners* (Oxford University Press, 1975)

4 Harvey, D., *Social Justice and the City* (Edward Arnold, 1973). This is an essential starting-point for any attempt to understand geography's transitions since the Quantitative Revolution and its formalisation in Harvey's earlier *Explanation in Geography* (Edward Arnold, 1969)

5 Bunge, W., *Theoretical Geography* (Gleerup, 1962); for a spirited counter-attack, see Sack, R. D., 'The spatial separatist theme in geography', *Economic Geography* **50** (1974), 1–19

6 I have provided a fuller discussion in Gregory, D., *Social Theory and Spatial Structure* (Hutchinson, 1982), and much of the argument which follows is developed in greater detail there

7 Thrift, N., 'Social theory and human geography', *Area* **13** (1981)

8 Johnston, R. J., *op. cit.*

9 I think this was particularly true of many of the contributions to *Antipode: a radical journal of geography* which, with some notable exceptions, ignored developments in modern European Marxism: see Anderson, P., *Considerations on Western Marxism* (New Left Books, 1976)

10 See, for example, Muir, R., 'Radical geography or a new orthodoxy', *Area* **10** (1978), 322–7; Walmsley, D. J., Sorenson, A. D., 'What Marx for the radicals?' *loc. cit.* **12** (1980), 137–41

11 I have discussed some of the connections between humanism in geography and Marxian humanism in Gregory, D., 'Human agency and human geography', *Trans. Inst. Brit. Geog.* **6** (1981), 1–18

12 BARNBROCK, J., 'Prolegomenon to a methodological debate on location theory: the case of von Thünen', *Antipode* **6** (1974), 59–66; Gregory, D., 'Alfred Weber and location theory', in Stoddart, D. R. (ed.), *op. cit.*; Entrikin, J. N., 'Robert Park's Human Ecology and Human Geography', *Ann. Assoc. Am. Geog.* **70** (1980), 43–58

13 See, for example, Billinge, M., Gregory, D., Martin, R. L., (eds), *Geography As Spatial Science: recollections of a revolution* (Macmillan, 1982); Freeman, T. W., *A History of Modern British Geography* (Longman, 1980); Johnston, R. J., *op. cit.*; Stoddart, D. R. (ed.), *op. cit.*

14 The most sensitive discussion of these issues is probably still Buttimer, A., 'Values in geography', Assoc. Am. Geog. Commission on College Geography, *Resource Papers*, **24** (1974); for an attempt to connect these to modern critical theory – and to the work of Jürgen Habermas in particular – see Gregory, D., *Ideology, Science and Human Geography* (Hutchinson, 1978). Substantive surveys which acknowledge the central importance of politics and ethics include Coates, B. E., Johnston, R. J., Knox, P. L., *Geography and Inequality* (Oxford University Press, 1977); Smith, D. M., *Human Geography: a welfare approach* (Edward Arnold, 1977); idem., *Where the Grass is Greener* (Penguin, 1979)

15 LEY, D., 'Geography without man: a humanistic critique', School of Geography, University of Oxford, *Research Papers* **24** (1980). Harvey, D., 'Population, resources and the ideology of science, *Econ. Geog.* **50** (1974) 256–77

16 The most sophisticated discussion of systems theory in modern geography is Bennett, R. J., Chorley, R. J., *Environmental Systems: philosophy, analysis and control* (Methuen, 1978), while structuralism is discussed in Gregory, D., 'Ideology' *op. cit.*, Ch. 3

17 For a review, see Gregory, D., 'Ideology' *op. cit.*, Ch. 4; see also Ley, D., Samuels, M. (eds), *Humanistic Geography: prospects and problems* (Croom Helm, 1978)

18 The theory of structuration has been developed by a host of social theorists, but the version set out here is drawn from A. Giddens, *New*

Rules of Sociological Method (Hutchinson, 1976), and idem., *Central Problems in Social Theory* (Hutchinson, 1979)

19 *Ibid.*

20 CURRY, L., 'Position, flow and person in theoretical economic geography', in Carlstein, T., Parkes, D., Thrift, N. (eds), *Timing Space and Spacing Time:* vol. 3: *Time and Regional Dynamics* (Edward Arnold, 1978)

21 OLSSON, G., *Birds in Egg/Eggs in Bird* (Pion, 1980)

22 CLARKSON, J. D., 'Ecology and spatial analysis', *Ann. Assoc. Am. Geog.* **60** (1976) 700–16

23 HURST, M. E., 'Geography, social science and society: towards a definition', *Australian Geog. Studs.* **18** (1980), 3–20; his *objection* is correct, but the terms in which it is lodged are, I think, much more contentious

24 See Schmidt, A., *The Concept of Nature in Marx* (New Left Books, 1971)

25 SAYER, R. A., 'Epistemology and conceptions of people and nature in geography', *Geoforum* **10** (1979), 19–44; see also Burgess, R., 'The concept of nature in geography and Marxism', *Antipode* **10** (1978), 1–10, and Anderson, J., 'Towards a materialist conception of geography', *Geoforum* **11** (1980), 171–8

26 CHORLEY, R. J., 'The role and relations of physical geography', *Progress in Geography* **3** (1971), 87–109; Gregory, K. J., Walling, D. E. (eds) *Man and environmental processes* (Dawson, 1979); see also Bennett, R. J., Chorley, R. J., *op. cit.*

27 Much of modern economic geography is drawn to a materialist philosophy and a structural method, while cultural geography often endorses an idealist philosophy and an interpretative method; stark though these divisions are, they are by no means *necessary* ones

28 OLWIG, K., 'Historical geography and the society/nature problematic', *J. Hist. Geog.* **6** (1980) 29–45. There have been a host of studies of environmental imagery – see, for example, Yi-Fu Tuan, *Topophilia* (Prentice-Hall, 1974) – and these can be connected to the next section through parallel studies of spatial imagery – see, for example, Sack, R. D., *Conceptions of Space in Social Thought* (Macmillan, 1980); but both sorts of study need to specify the relations between these cultural constructions and the conduct of practical life

29 SOJA, E. 'The socio-spatial dialectic', *Ann. Assoc. Am. Geog.* **70** (1980), 207–25

30 For introductions to time-geography, see Carlstein, T., Parkes, D., Thrift, N. (eds), *op. cit.*, vol. 2: *Human Activity and Time Geography* (Edward Arnold, 1978); Parkes, D., Thrift, N., *Times, Spaces and Places* (Wiley, 1980)

31 HAGERSTRAND, T., 'The domain of human geography', in Chorley R. J. (ed.), *Directions in Geography* (Methuen, 1973)

32 But see Pred, A., 'Social reproduction and the time-geography of everyday life', *Geografiska Annaler* **63B** (1981), 5-22.

33 HOLLAND, S., *Capital Versus the Regions* (Macmillan, 1976)

34 CLARK, G., 'Capitalism and regional inequality', *Ann. Assoc. Am. Geog.* **70** (1980), 226-37; see also Dunford, M., 'Capital accumulation and regional development in France', *Geoforum* **10** (1978), 81-108

35 See, for example, Dear, M., Clark, G. L., 'The state and geographic process: a critical review', *Environment and Planning* **10**A (1978), 173-84; *idem*, 'The state in capitalism and the capitalist state', in Dear, M., Scott, A. (eds), *Urbanization and Urban Planning in Capitalist Societies* (Methuen, 1981); Johnston, R. J., 'Political geography without politics', *Progress in Human Geography* **4** (1980), 439-46

36 See, for example, Bennett, R. J., *The Geography of Public Finance* (Methuen, 1980)

37 See, for example, Martin, R. L., *Wage Inflation in Urban Labour Markets* (Methuen, in press)

38 See, for example, Massey, D., 'Capital and locational change: the U.K. electrical engineering and electronics industries', *Rev. Rad. Pol. Ec.* **10** (1978), 39-54; *idem.*, Meegan, R. A., 'The geography of industrial restructuring', *Progress in Planning* **10** (1979), 155-37; Thrift, N., 'Unemployment in the Inner City: Urban problem or structural imperative?', in Herbert, D. T., Johnston, R. J. (eds), *Geography and the Urban Environment*, vol. 2 (Wiley, 1979)

39 See, for example, Bassett, K., Short, J., *Housing and residential structure: alternative approaches* (RKP, 1980); Harvey, D., 'Class structure in a capitalist society and the theory of residential differentiation', in Peel, R., Haggett, P., Chisholm, M. (eds), *Processes in Physical and Human Geography: Bristol essays* (Heinemann); Scott, A. J., *The Urban Nexus and the State* (Pion, 1980); for an introductory review of several of the issues raised in these outstanding contributions, see Johnston, R. J., *City and Society: an outline for urban geography* (Penguin, 1980)

40 BLAIKIE, P., 'The theory of the spatial diffusion of innovations: a spacious cul-de-sac', *Progress in Human Geography* **2** (1978), 268-95; Yapa, L. S., 'The Green Revolution: a diffusion model', *Ann. Assoc. Am. Geog.* **67** (1977), 350-9

41 HARVEY, D., 'The geography of capitalist accumulation: a reconstruction of the Marxian theory', in Peet, R. (ed.), *Radical Geography* (Methuen 1978)

42 HAGGETT, P., *Locational Analysis in Human Geography* (Edward Arnold, 1965)

43 WALKER, R., 'Two sources of uneven development under advanced capitalism: spatial differentiation and capital mobility', *Rev. Rad. Pol. Ec.* **10** (1978), 28-37

44 An obvious example of this reduction and redefinition is Chisholm, M., *Human Geography: Evolution or Revolution?* (Penguin, 1975).
45 GREGORY, D., 'Ideology', *op. cit.*

8 Geography and values education

John Huckle Bedford College of Higher Education

> 'If I had my chance now, in terms of a curriculum project for the seventies, it would be to find a means whereby we could bring society back to its sense of values and priorities in life.'
> JEROME BRUNER[1]

By the time *New Directions in Geography Teaching* was published in 1973, at least one founding father of the curriculum reform movement had become disenchanted with its early products. Geographical education in Britain was however to overlook such warnings, and enjoy a romance with science and logical curriculum planning which fundamentally changed the professional outlooks of geography teachers. This chapter will suggest that many of these teachers now enter the 1980s in much the same spirit as that expressed by Bruner in 1971. They are aware of the gains brought by the reformed geography of the 1970s, but sense that we must now restore the balance between the cognitive and affective in geography classrooms. Before outlining some developments which can assist us in this task, it is necessary to examine the origins of our new concern for values.

Value trends in the 1970s

The decline of traditional cultural and religious values is a long established trend in society related to the far-reaching processes of modernisation. Changes in economic and social structure, often justified now in terms of the requirements of technology and bureaucracy, cause modern man to adopt a belief system which is more concerned with means than ends, and promotes such instrumental values as efficiency, materialism and individualism, at the expense of such absolute values as human dignity, social justice and environmental well being. Our dominant style of thought has become so objective and reductionist that it often causes us to overlook the quality of human and man–environment relationships, while our political awareness has been so deadened that we fail to recognise situations in which economic interests now dominate over human needs. The neglect of feelings and emotions in a cultural climate of political apathy and pragmatism causes us to become estranged from ourselves and the world in which we live. Once lost and alienated, many are seduced into the belief that values are just a matter of opinion and retreat to a state of moral relativism or cultural nihilism[2] – 'it's good if I like it, if it's novel, or if it pays'. Crawford summarises the sense of loss which prompts others to oppose such cultural trends:

> '... the loss of a sense of dignity and significance of self and others, the loss of an ability to communicate deeply and personally with one another, the loss of relatedness to nature ... we have lost the sense of the tragic significance of life.'
> TERRAYNE CRAWFORD[3]

During the late 1960s and throughout the 1970s, this opposition found expression in both popular and academic movements which challenged the central values of industrial society. The optimism of the 1960s gradually changed to pessimism, with the growing realisation that man on 'spaceship earth' faced mounting problems of environmental, economic and political instability. For some, the increasing personal, social and environmental costs of technological progress were symptoms of a crisis in personal consciousness and lifestyles,[4] while for others, they were the related symptoms of a predicted crisis in advanced capitalism.[5] A counter-culture of hippies, mystics and communards, attracted the many idealists who sought subjectivity, spontaneity and community in alternative life-styles, while various left-wing groups catered for radicals who demanded a more active role in overthrowing the *status quo*. Environmentalism and women's liberation are examples of social movements which have drawn on such utopian and radical thought in the past fifteen years, and aspects of their thought and behaviour are now widely diffused in society. Education did not escape the attention of libertarians and radicals in the 1970s. Renewed attention was given to notions of progressive and community education, free schooling and de-schooling were hotly debated, and the radical school movement[6] alerted us to the de-humanising effects and social control functions of much that passed for teaching and learning.

Academic communities in the 1970s were also characterised by self-doubt and fierce debates on social responsibility. The rationalism which dominated geography and curriculum studies at the start of the decade was gradually challenged by humanists who perceived their new or rediscovered approaches to be of greater relevance to man's true needs, and the reform of an unjust world. While behavioural, humanistic and welfare geographies posed liberal or reformist challenges to the subject's new establishment, radical geography offered a more far-reaching critique by emphasising the ideological nature of much that was now accepted as geographical science.[7] Similarly, phenomenological and neo-Marxist approaches in education challenged the new tenets of curriculum planning by stressing the relativity of school knowledge, and initiating a vigorous debate on the role of schooling in maintaining and reproducing the existing social order.[8] Both geography and curriculum studies became characterised by a plurality of viewpoints and the task of designing a school curriculum became ever more problematic. Above all, it was no longer possible to ignore values.

Values and schooling

'We don't need no education,
We don't need no thought control,

Hey teacher, leave those kids alone,
All in all, you're just another brick in the wall.'
PINK FLOYD, 1980

The sounds I heard on the car radio while travelling home from Charney Manor in 1980 remind us of an adolescent culture which expresses the impossibility of establishing meaning in a world which increasingly denies its existence. Confused values, abundant choice and worsening social problems challenge schools to assist pupils in rediscovering and recreating absolute values, and in discarding those dysfunctional norms which currently contribute to relativism and nihilism. As geography teachers, we must seriously question whether our curriculum is just 'another brick in the wall', or whether it enables pupils to derive, clarify and apply values in such a way as to counter confusion, guide choice and contribute to the solution of problems which threaten our very survival. In a multi-ethnic society with high levels of youth unemployment, the plight of many youngsters is not unrelated to that of others in places near and far. A relevant school geography will not only offer a range of explanations of such an unjust world, but will provide pupils with some prospect of its reform.

The positions which schools and teachers currently adopt with regard to values education are determined by the educational ideologies (Ch. 14) which legitimate their activity. While a *conservative* ideology leads to attempts to inculcate 'right' values via approaches and rituals which are rarely free from elements of indoctrination, a more liberal approach employs the risk of indoctrination to justify a 'neutral' curriculum. *Liberals* see education as a neutral instrument of social policy and claim that it can pursue personal, social and economic aims without irreconcilable tensions developing. *Progressive*, or *child-centred*, teachers are more prepared to foster affectivity, but their preoccupation with personal morality, aesthetics and creativity means that lessons often fail to connect with issues in the real world. During the 1970s, an interpretative sociology of education revealed the differing ways in which conservative, liberal and progressive ideologies sustain prevailing patterns of injustice, and enable schools to reproduce inequality.[9] It prompted much critical evaluation of existing overt and hidden curricula, and focused attention on *radical* or *reconstructional* ideology which, by employing recent advances in moral and political education, enables a more justifiable approach to values education in schools.

Reconstructionism stems from a conviction that all is not well with the world, and that education should be a force for planned democratic change within the community. Teachers are to adopt a critical attitude towards contemporary social change, and make their curriculum relevant to the wider debate on alternative futures. By developing rationality, problem-solving ability, sensitivity and feeling, they are to cultivate autonomous individuals who are capable of making their own judgements and defending them rationally, emotionally and aesthetically. Autonomy, in such areas as morality and politics, provides a ready defence against the frequent charge of indoctrination which reconstructionists face, and ensures that pupils can recognise and attack bias wherever it is experienced. Reconstructionism requires a commitment to social justice and the community, and a great respect for procedural openness and creativity in the classroom. While its literature is now extensive,[10] there appear to have been few geography teachers amongst its recent advocates. Our failure to engage and develop such viewpoints is perhaps a symptom of other preoccupations during recent years.

Values in geographical education

At Charney Manor in 1970, there was clearly an acknowledgement of affective objectives, and the fact that many issues dealt with in the geography curriculum could engage the pupil's values. Geography's contribution to citizenship education was discussed, and both role play and decision-making exercises were mentioned in the context of affective outcomes. The papers from the conference[11] do however reflect a genuine uncertainty about values education and the teacher's responsibility in this area:

> 'The aims implicit in the affective domain receive only lip service in geography.'
>
> DAVID GOWING, p. 155

> 'At present our work rarely demands attitudes, values, and emotions.'
>
> DAVID GOWING, p. 156

> 'The various objectives I have linked together under the heading of citizenship present very considerable problems, mainly because they are concerned with attitudes as much as with understanding, and we

know very little about the type of teaching most likely to bring about adherence to value systems and forms of behaviour that we may strongly approve of'

TREVOR BENNETTS, p. 170

'The problems of the desirability or otherwise of teaching attitudes and values ... are not easy to resolve.'

REX BEDDIS, p. 179

Reform of geographical education in the 1970s was essentially a reform of its cognitive curriculum in which values were overlooked to a considerable extent. In 1970, curriculum developers were preoccupied with new ideas, theories and skills, and the rational planning models they employed reinforced the separation of knowledge and values implicit within their new content. The reformed curriculum fitted well with the needs of a technological society and teacher professionalism,[12] and although the two 14–16 Schools Council projects recognised values and attitudes, they appear to have been of secondary importance from the start, and were further relegated as teachers adapted the materials to their own classroom needs.[13] While geographers were engaged in reform, others were planning and implementing courses which may be seen as more realistic and responsible reactions to mounting social and environmental problems.[14]

By the end of the decade, there were signs that the renewed attention to values in academic geography had prompted a recognition of their neglect in schools. While some[15] continued the debate on a suitable rationale and approach, others[16] provided clear evidence that considered approaches existed and merely needed wider dissemination. A consideration of these approaches must begin with an account of the function of values within the individual's personality.

Psychological approaches to the person

If it is the pupil's affective life which connects him to the facts and ideas presented in geography lessons, the teacher must base his pedagogy on models of the person which explain the interaction of thought and feelings, and allow for such differing dimensions of affect as attitudes, values, and emotions. Figure 26 represents one such model of the way in which the individual's behaviour is shaped by his beliefs and values. At the 'heart' of the individual are his terminal values, the fundamental principles such as social justice or environmental well being on which he is rarely willing to compromise. He also holds instrumental values which suggest appropriate modes of conduct to attain these terminal goals. Terminal and instrumental values are linked in complex ways and serve to develop and maintain the attitudes which govern our day-to-day behaviour. Faced

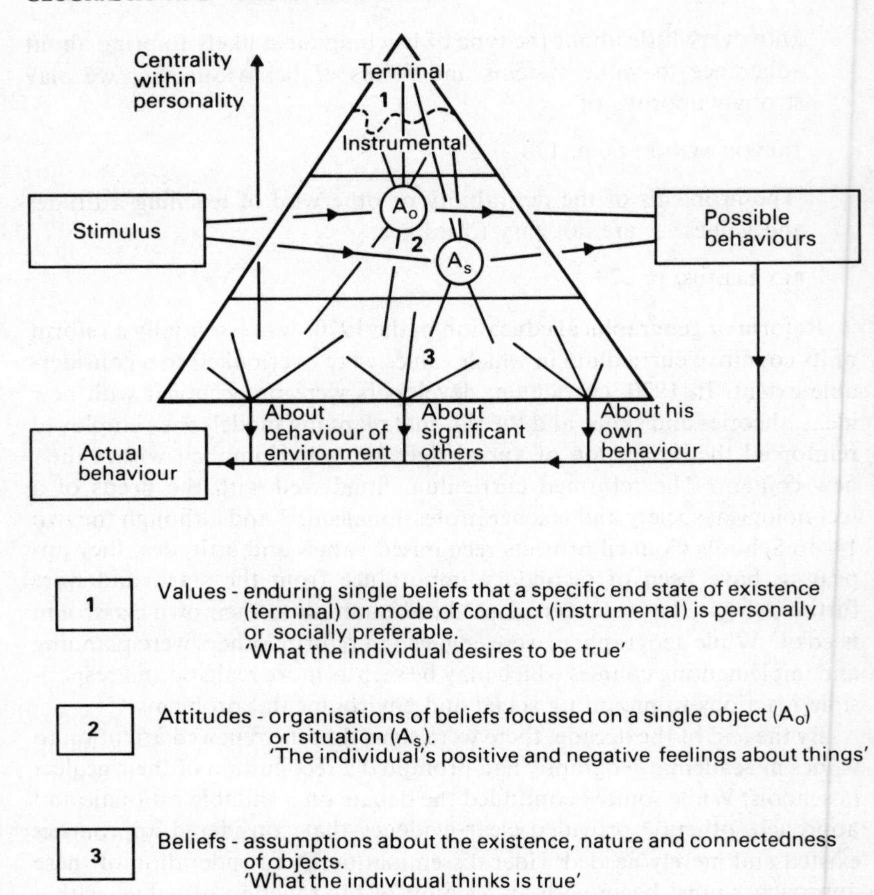

Fig. 26 *The way in which the individual's behaviour is shaped by his beliefs and values (based on Rockeach*[17])

with a novel situation, we draw upon both our beliefs and values to form attitudes towards the stimulus object (A_o in Fig. 26), and the situation in which it occurs (A_s). Together these suggest possible behaviours which are assessed in the light of other beliefs to determine their feasibility and acceptability. Our actual behaviour is then the outcome of cognitive and affective elements of our personality interacting in complex ways. Much of it is routine behaviour, in which the above process is not fully activated, and conscious moral judgement is not required. Values education based on this model would aim for greater coherence within the pupil's belief system, and would help him to recognise and remedy inconsistencies between the principles it generates and his own behaviour. It would ask him to examine and reconcile knowledge and values with respect to controversial issues and so generate moral decisions and action. Such assumptions are strongly reflected in the values analysis approaches considered below.

The model just described adopts a scientific approach to the person in that it reduces man to psychological constructs, and suggests that he strives to behave in a rational and predictable way. If we know enough about his values, beliefs and environment, we can predict his attitudes and resulting behaviour. Alternative conceptions of man's affective life are provided by such humanistic psychologists as Rogers and Maslow, who stress that man is essentially arational, unpredictable and free to choose within the parameters of his own consciousness and experience. They emphasise that the individual lives in many lifeworlds, each governed by a distinct sphere of consciousness with its own meanings. The pupil at school, at home or at a disco, is essentially different people governed by different sets of beliefs. This phenomenological view of man stresses his need for positive self regard and the role of subjective experience in enabling self actualisation. 'Right' behaviour is generated through an understanding of self and situation which results not from a rational consideration of values and beliefs but from an openness to the meanings of life's encounters. Wonder, admiration, contemplation and celebration are the state we should encourage our pupils to attain if they are to become the fully human, feeling and creative individuals we require. Humanistic educators[18] have developed approaches which emphasise such unity of consciousness or balance between heart and mind. Intuition and imagination are to complement rationality, and techniques of consciousness expansion are to find a place within the classroom.

Approaches to values education

Methods of handling values in the classroom are then guided by different conceptions of education and of man's affective life. The reconstructionist must select approaches which foster autonomy and are based on models of the valuing process which allow uncoerced choice. These approaches allow the teacher to become the facilitator rather than the arbitrator of value decisions, and may be matched to the range of objectives which values education subsumes in the manner shown by Table 15. Heavily based on the work of Superka et al.,[19] this table attempts to summarise the purposes and methods of each approach, and suggests texts which provide a suitable introduction to its literature. The reader should recognise that the allocation of some texts to particular categories is somewhat arbitrary, and that other references are needed to relate the approaches to geographical education.[20] These make the important points that few moral or social issues can be resolved within the constraints of one discipline, and that values education should generally be regarded as an interdisciplinary activity. A discussion of the four major approaches follows:

Table 15. *Values education approaches*

Approach	*Purposes*	*Methods*	*Useful texts*	
Analysis	To help students use logical thinking and scientific investigation to decide value issues and questions To help students use rational, analytical processes in interrelating and conceptualising their values	Structured rational discussion that demands application of reasons as well as evidence; testing principles; analysing analogous cases; debate; research	*Values Education* *Practical Methods of Moral Education* *Political Education and Political Literacy* *Debate and Decision*	Metcalf, 1971 Wilson, 1972 Crick & Porter, 1978 World Studies Project, 1980
Moral development	To help students develop more complex moral reasoning patterns based on a higher set of values To urge students to discuss the reasons for their value choices and positions, not merely to share with others, but to foster change in the stages of reasoning of students	Moral dilemma episodes with small-group discussion relatively structured and argumentative	*Startline* and *Lifeline* *Teaching for Empathy* *Promoting Moral Growth*	Schools Council/Longman Schools Council/Collins 1976 Hersh et al, 1979

Clarification	To help students become aware of and identify their own values and those of others To help students communicate openly and honestly with others about their values To help students use both rational thinking and emotional awareness to examine their personal feelings, values and behaviour patterns	Role-playing games; simulations; contrived or real value-laden situations; in-depth self-analysis exercises; sensitivity activities out-of-class activities; small group discussion	*Values and Teaching* *Values Clarification* *Advanced Value Clarification* *Meeting yourself half-way*	Raths, 1966 Simon et al 1972 Kirschenbaum, 1977 Simon, 1974
Action learning	These purposes listed for analysis and clarification To provide students with opportunities for personal and social action based on their values To encourage students to view themselves as personal–social interactive beings, not fully autonomous, but members of a community or social system	The methods listed for analysis and clarification as well as action projects within the school and community and skill practice in group organising and interpersonal relations	*Organising Community Service* *Education for a Change* *Community Service in Education* *Ideas into Action*	C.S.V., 1972 Ball & Ball, 1973 D.E.S., 1974 World Studies Project, 1980

Based on Superka et al., 1976 (19)

1 Values analysis

This approach to values education emphasises that moral judgements are based on facts and values, and that reason is the foundation of values analysis in education. The role of the teacher is to teach the decision-making process while leaving the pupil free to arrive at his own independent judgement. Faced with an issue such as the siting of a reservoir, or the suitability of an aid programme, the pupil is to assess the factual evidence he considers relevant in the light of his values, and arrive at a tentative judgement. By studying relevant knowledge, clarifying appropriate values, and assessing the possible consequences of alternative policies, he is to engage, at his own level, in the type of decision making carried out by planners and others. This approach fits well with recent reforms in geographical education, for the pupil is to generate value principles by methods which resemble hypothesis testing.

Tentative judgements are refined by subjecting them to tests of logical and ethical acceptability; processes in which group discussion plays a crucial role. While decision-making exercises, often in the form of role play, have become a familiar element of geography teaching in the past decade, the literature suggests a low state of awareness of values education developments within the American social studies curriculum.

A review of the American literature[21] reveals that approaches to values analysis are more thoroughly developed than in Britain, since there has been greater attention to those cognitive and affective abilities which contribute to sound decision making and moral autonomy. A range of process models are available, together with curriculum materials which give equal attention to knowledge and values. The '16–19' project in geography has clearly benefited from a limited exposure to this literature, and its route for enquiry (similar to Banks's value inquiry model shown as Fig. 27) promises to reinforce a cautious move into values analysis begun by the two earlier projects. Teachers wishing for a more accessible source of insights than the American texts should talk to colleagues teaching moral education. John Wilson's work attempts to demonstrate how the components of moral autonomy may be taught and assessed in the classroom, and I have modified these elsewhere in discussing the cultivation of an environmental ethic within geographical education.[22]

The major weaknesses of values analysis relate to its readiness to cling to what some consider a debased form of rationality, and to continue to promote knowledge above values. Inglis[23] accuses such curriculum buildings as the geography project teams of promoting ideas and cognitive skills at the expense of values and feelings, and so relegating these to an afterthought. He suggests that approaches to rational decision making in the classroom are often so dominated by the desire for balance and consensus, that rationality becomes mere judicious moderation. In the search for consensus, respect for absolute truth and values is likely to be com-

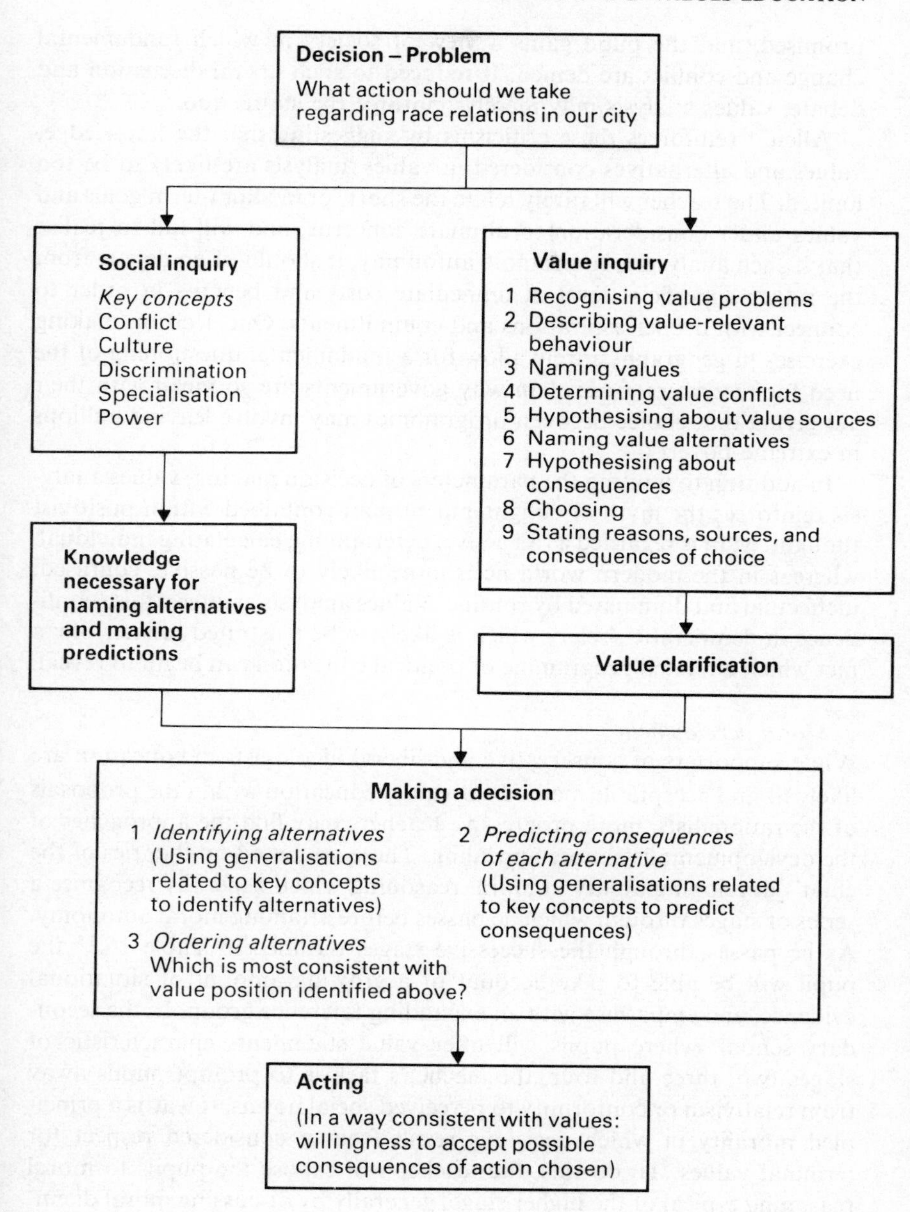

Fig. 27 *The decision-making process* (Banks and Clegg, *Teaching Strategies for the Social Studies: Inquiry, Valuing and Decision Making*, Addison-Wesley Publishing Company Inc, 1977)

promised, and the pupil gains a view of society in which fundamental change and conflict are denied. If reduced to such liberal discussion and debate, values analysis may merely reinforce the status quo.

Allen[24] reinforces these criticisms by suggesting that the knowledge, values and alternatives considered in values analysis are likely to be too limited. The teacher will rarely relate the short- or medium-term goals and values under consideration to ultimate concerns, and will fail to realise that if such analysis is to promote autonomy, it should often depart from the rather superficial level of immediate costs and benefits in order to connect with deeper aspirations and commitments. Our decision-making exercises in geography rarely allow for a fundamental questioning of the need for reservoirs, or explain why governments are so mean with their aid giving that choice between programmes may involve leaving millions in extreme poverty.

In addition to limiting the parameters of decision making, values analysis reinforces the myth of enlightenment man contained within positivist thought. Man is regarded as an active, determining, calculating individual, whereas in the modern world he is more likely to be passive, confused, ineffectual and dominated by routine. Values analysis creates a false confidence in democratic society which is likely to be frustrated in later life; a fact which a related programme of political education can begin to reveal.

2 Moral development

While supporters of conservative and liberal ideologists in education are likely to find acceptable methods of values education within the proposals of the rationalists, more progressive teachers may find the approaches of the developmentalists more appealing. These are based on theories of the child's developing ability in moral reasoning which generally recognise a series of stages through which he passes before attaining moral autonomy. As he passes through the successive stages outlined in Table 16,[25] the pupil will be able to take account of a growing amount of situational evidence, and empathise with an expanding reference group. In the secondary school, where pupils will offer value statements characteristic of stages two, three and four, the teacher's task is to prompt pupils away from relativism or conformity to perceived social norms, towards a principled morality in which decisions result from a considered respect for terminal values. To do this, the teacher will expose the pupils to moral reasoning typical of the higher stage, generally by discussing moral dilemmas in small group situations. Such discussion requires pupils to acknowledge the nature of their reasoning and its deficiencies, and provides an opportunity for exposing them to more autonomous thought. The texts listed in Table 15 give advice on both the construction and presentation of moral dilemmas in the classroom, and many of the case studies used in geography provide suitable content for adaptation.

Table 16 *Definition of moral stages*

I Preconventional level

At this level, the child is responsive to cultural rules and labels of good and bad, right or wrong, but interprets these labels either in terms of the physical or the hedonistic consequences of action or in terms of the physical power of those who enunciate the rules and labels.

Stage 1: The punishment and obedience orientation. The physical consequences of action determine its goodness or badness, regardless of the human meaning or value of these consequences.

Stage 2: The instrumental–relativist orientation. Right action consists of that which instrumentally satisfies one's own needs and occasionally the needs of others.

II Conventional level

At this level, maintaining the expectations of the individual's family, group or nation is perceived as valuable in its own right, regardless of immediate and obvious consequences.

Stage 3: The interpersonal concordance of 'good boy–nice girl' orientation. Good behaviour is that which pleases or helps others and is approved by them.

Stage 4: The 'law and order' orientation. There is orientation toward authority, fixed rules and the maintenance of the social order.

III Postconventional, autonomous or principled level

At this level, there is a clear effort to define moral values and principles that have validity and application apart from the authority of the groups of persons holding these principles and apart from the individual's own identification with these groups.

Stage 5: The social - contract, legalistic orientation, generally with utilitarian overtones. Right action tends to be defined in terms of general individual rights and standards which have been critically examined and agreed upon by the whole society.

Stage 6: The universal ethical principle orientation. Right is defined by the decision of conscience in accord with self-chosen ethical principles appealing to logical comprehensiveness, universality and consistency.

KOHLBERG[25]

3 Values clarification

A third approach to values education, which is very popular in America, is values clarification. This is based on a simple model of the valuing process which suggests that valuing involves free choice from considered alternatives, and that having made a choice, the pupil should be pleased to both publicly affirm it, and repeatedly act in accordance with his decision. The techniques developed to enable pupils to act out their choices, affirmations and behaviours in the classroom are very novel, and have a strong appeal to the teacher seeking ready-made exercises. Some are best regarded as starters or summaries to more lengthy pieces of work, for the main weakness of the approach is that it can encourage moral relativism by suggesting that one's values are one's own concern and that, provided one is happy with one's choice, all is well. Values education will generally go beyond awareness and clarification, and this approach will need to be complemented by others. Examples of values clarification in geography are not that plentiful,[26] but at least one curriculum project in environmental education has employed the approach.[27]

Aesthetic support. While affective education draws upon all the approaches considered so far, we have yet to identify approaches which fully reflect the phenomenological view of man outlined above. There can be little doubt that recent changes in academic geography have caused us to re-examine the manner in which we experience place. That contact with local and distant environments and their inhabitants can lead to consciousness expansion is a phenomena of which we have long been aware, but the cultivation of the affective aspects of such experience in our classrooms must be rediscovered and developed. If school geography is to foster intuition, fantasy and feeling, then we will need to make far more use of aesthetic support to our teaching.[28] Music, art and literature of all kinds must be used to evoke the subjective meanings of place and counter the cold objectivity of much that currently takes place in classrooms. Teachers who have come through the reforms of the past decade may find the prospect of a humanistic revolution rather daunting, but the discovery of landscape through painting, poetry and music should be more enjoyable to some than was its investigation via spatial analysis and statistics.

Again American educators appear to have gone further than their British counterparts in applying humanistic psychology to the classroom. British readers may find some discussions of humanistic and transpersonal education[29] so far removed from current practice in schools as to be irrelevant, but it is likely that such theories of education provide the frameworks into which the new humanistic geography can be absorbed. It is fascinating to speculate that geographers could at last find a philosophical rationale for the humanities, and help create an integrated curriculum which would function as a true vehicle for self discovery.

Political education. If values education is to contribute to both the development of the individual and the reshaping of society, the pupil must be shown how his individual decision and commitment to action can find expression in the wider community. Fortunately the past decade has seen the reform of political education which has now rejected approaches to the teaching of civics and government based on the inculcation of values, in favour of the pursuit of political literacy. The Programme for Political Literacy has carried out an analysis of the knowledge, skills and values needed by the individual who is capable of understanding and participating in political issues, and its classroom approaches may be seen as a necessary extension to values analysis.

> 'A politically literate person will then know what the main political disputes are about; what beliefs the main contestants have of them; how they are likely to affect him, and he will have a predisposition to try to do something about it in a manner at once effective and respectful of the sincerity of others.'
>
> BERNARD CRICK AND IAN LISTER[30]

Many topics in geography have a political dimension in that they can only be understood by reference to the processes which determine the distribution of power and resources in society. If the pupil is to recognise such political dimensions, and contemplate strategies of influence in a rational and tolerant way, the teacher will need to make use of such checklists as those outlined by Stradling.[31] These emphasise the scope for values analysis and clarification within political education, and alert the teacher to the wide range of relevant knowledge, skills and attitudes and values required. Whether the dispute is about North Sea fisheries, boundaries within the Middle East or the local council's reluctance to reclaim derelict land, this checklist suggests what is minimally necessary before a class can realistically discuss the issue. In addition to such checklists, the Programme for Political Literacy provides an account of the key political concepts such as natural rights, welfare and justice, which should be constant reference points in political education. While the output of welfare and radical geographers has clearly added a new political dimension to the subject, due attention to the tenets of political education should ensure that their methodology and content can be adapted to the classroom in a defensible way.

While political education has not gone unnoticed by geographers, it has perhaps had a greater impact in the general field of environmental education[32] where planners have acknowledged it as a vehicle for fostering greater public participation, and certain Urban Studies Centres have combined it with community action.[33]

4 Action learning

Our success with values education will ultimately be judged by the behaviour of our pupils and their contributions to social reconstruction. Since both teachers and pupils are trapped within prevailing social norms and habit patterns, they may find it very difficult to change their behaviour despite a disparity between their declared and lived values. In such a situation, the teacher should try to show his pupils examples of behaviour which are more in keeping with his value principles, and help him to recognise the social and environmental constraints which prevent him translating concern into action. While there is frequently much scope within the school for creating a more just and caring community, the geographer will need to search for community service projects which take him and his pupils into the wider environment. Work with a local conservation corps or for the local Oxfam shop are concrete ways of improving the environment and the plight of the poor overseas, while a half day set aside for litter collection would be one way of ensuring that a field party contributed to the aesthetics of its host environment. Some would regard such action as rather distant from the realities of power, and would suggest that older pupils, who are particularly incensed about an issue, should become more directly involved in the political process. The teacher considering overtly political action should remember that schools are communities of children whose parent's wishes should be respected, and that there are generally sufficient opportunities to engage in responsible action without attracting accusation of professional irresponsibility.

Tasks for the 1980s

> The teaching of geography must thus pursue a treble aim; it must awaken in our children the taste for natural science altogether; it must teach them that all men are brethren, whatever be their nationality; and it must teach them to respect the 'lower races'. Thus understood, the reform of geographical education is immense: it is nothing less than a complete reform of the whole system of teaching in our schools.
>
> PETER KROPOTKIN[34]

There has been a considerable progress in geographical education since Kropotkin issued his challenge almost 100 years ago. The present need is to acknowledge the costs and benefits of recent reform, and realise that newly acquired ideas and skills must be supplemented by values education if our classrooms activity is to contribute to the pupil's sense of identity, and greater justice in the world in which we live. Equipped with a wider range of geographies and approaches to curriculum planning, we are in a strong position to enter a new phase of curriculum development. The initial and in-service education of geography teachers in the 1980s must regard values education as a priority, and we must establish a dialogue with moral and political educators which prompts extra effort and resources being devoted to the development of suitable classroom activities. The task is at least as demanding as that which was discussed at Charney Manor in 1970, but the potential rewards? . . . they, as both Kropotkin and Bruner realised, are far greater.

References

1 BRUNER, J. S. (1971) *The Process of Education Revisited.* Phi Delta Kappan, September

2 HOLBROOK, D. (1977) *Education, Nihilism, and Survival.* Darton, Longman & Todd

3 CRAWFORD, T. (1975) 'Values and the educational process', in Haavelsrud, M. (ed.) *Education for Peace.* IPC

4 ROSZAK, T. (1970) *The Making of a Counter Culture.* Faber & Faber

5 HABERMAS, J. (1976) *Legitimation Crisis.* Heinemann

6 GOSS, R. and GOSS, B. (1972) *Radical School Reform.* Penguin
BARROW, R. (1978) *Radical Education.* Martin Robertson

7 SMITH, N. (1979) 'Geography, science and post positivist modes of explanation', *Progress in Human Geography*, **3,** 356–83

8 SARUP, M. (1978) *Marxism and Education.* Routledge and Kegan Paul

9 DALE, R. et al. (eds) (1976) *Schooling and Capitalism*, Routledge & Kegan Paul, for Open University Press.
YOUNG, M. and WHITTY, G. (eds) (1977) *Society, State, and Schooling.* Falmer Press.

10 WREN, B. (1977) *Education for Justice.* SCM Press. Lynch, J. (1979) *Education for Community.* Macmillan Education.

11 WALFORD, R. (ed.) (1973) *New Approaches to Geography Teaching.* Longman

12 HUCKLE, J. (ed.) (forthcoming) *Geographical Education: reflection and action.* OUP

13 HUCKLE, J. (1976) 'Geography for the young school leaver – a classroom opener?', *Classroom Geographer*, January

14 WORLD STUDIES PROJECT (1976) *Learning for Change in World Society* One World Trust
RICHARDSON, R. (1977) *World Studies* (a textbook series). Nelson

15 COWIE, P. M. Value Teaching and Geographical Education, unpublished M.A. dissertation, University of London
WATSON, J. W. (1977) 'On the teaching of value geography', *Geography*, **62** (276), July
SMITH, D. L. (1978) 'Values and the teaching of geography', *Geographical Education*, **3** (2)

16 ALLEN, R. F. (1975) *Teaching Guide for the Values Education Series.* McDougal, Little & Co.
MARTORELLA, P. H. (1977) 'Teaching geography through value strategies', in Manson G. A. & Ridd M. K. (eds) *New Perspectives on Geographic Education.* Kendall Hunt, for NCGE
HUCKLE, J. (1980) 'Values and the teaching of geography – towards a curriculum rationale', *Geographical Education*, **3,** (4)

17 ROKEACH, M. (1973) *The Nature of Human Values.* The Free Press, a division of Macmillan Publishing Co Inc.

18 MILLER, J. P. (1976) *Humanising the Classroom*. Praeger
19 SUPERKA, D. P. et al. (1976) *Values Education Sourcebook*, ERIC Clearing house for Social Science/Social Studies Education, Boulder, Colorado
20 ALLEN, R. F. and MARTORELLA, P. H. Ref. 16, *op. cit.*
21 OLINER, P. M. (1976) *Teaching Elementary Social Studies*. Harcourt Brace, Janovich
BANKS, J. A. (1977) *Teaching Strategies of the Social Studies*. Addison Wesley.
MASSIALAS, B. G. and HURST, J. B. (1978) *Social Studies in a New Era*, Longman.
22 HUCKLE, J. (1978) 'Geography and values in higher education', *Journal of Geography in Higher Education*, **2** (1), Spring.
23 INGLIS, F. (1974) 'Ideology and curriculum', *Journal of Curriculum Studies*, **6** (1)
24 ALLEN, R. F. Ref. 16, *op. cit.*
25 KOHLBERG, L. (1973) 'Stages of moral development', *The Journal of Philosophy*, 25 October
26 KRACHT, J. B. and BOEHM, R. G. (1975) 'Feelings about the community – using value clarification in and out of the classroom', *Journal of Geography*, **74** (4) April
27 BIOLOGICAL SCIENCES CURRICULUM STUDY (1975) *Investigating Your Environment*, Teacher's Handbook. Addison Wesley
28 READ, H. (1970) *The Redemption of the Robot*. Faber & Faber
29 ROBERTS, T. B. (ed.) (1975) *Four Psychologies Applied to Education*, Halstead Press
30 CRICK, B. and PORTER, A. (1978) *Political Education and Political Literacy*. Longman.
31 STRADLING, R. (1978) 'Political education in the 11–16 curriculum', *Cambridge Journal of Education*, **8** (2 & 3)
32 FYSON, A. (1977) *Change the Street*. OUP
33 WEBB, C. (1979) 'Urban studies and political education', *School and Community*, **23,** Spring.
34 KROPOTKIN, P. (1885) 'What geography ought to be', *The Nineteenth Century*, December

9 Progression in the geography curriculum

Trevor Bennetts HMI

During the last decade many geography teachers have been involved in the introduction of new courses or the substantial modification of old ones. Although the patterns which have emerged are varied, the general trend has been towards a greater emphasis on understanding ideas, on applying more sophisticated techniques of enquiry and on the consideration of social and environmental issues. Whilst in the 1960s geography teachers, in the main, attempted to accommodate new ideas and methods within existing course frameworks, in the 1970s many of them engaged in more radical restructuring of their teaching programme. In some schools the task of redesigning geography programmes has been complicated by the introduction, in the lower school, of some form of combined studies within which geography is a component. But even within single-subject courses restructuring has been widespread, with a conspicuous shift away from regionally based courses towards a systematic arrangement of topics. There is no reason to suppose that the processes which lead to the need for curricular changes will somehow come to an end. As geography teachers respond to the variety of influences which help to shape the curriculum, including the stream of educational opportunities flowing from their parent discipline, they continually face the task of adjusting their programmes and evaluating course structures. It is, therefore, important that teachers acquire the knowledge and skills required to plan and design effective courses.

This paper is concerned with progression, an element in the structure of courses which has received rather less attention than it deserves. Such neglect is understandable. Textbook writers are rarely explicit about the principles they use to select and organise content, and most teachers have probably relied more on tradition, experience and an intuitive feel for what is appropriate for each stage of a course, than on any explicit recognition of principles concerned with the process of learning. Even the Schools Council curriculum projects have made only a limited contribution to this problem, probably because the three major geography projects are each concerned with pupils of a limited age span. Not surprisingly, therefore, it is the History, Geography and Social Science 8–13 project which has given most attention in sequence and progression (Blyth et al[1]), although, from a geographical point of view its treatment appears to be of a rather general nature. However, the current interest in geographical ideas and skills provides an opportunity to improve the ways in which we structure progression in the learning of ideas in geography, and the extent to which we can identify principles which should guide our organisation of progression within geography or related courses and, as pupils work their way through the school system, from one course to another.

Principles to guide progression

It may first be helpful to distinguish between the ideas of sequence and progression. A content-based syllabus inevitably displays some sort of sequence, as one topic follows another, but such a syllabus does not necessarily reveal features designed to facilitate progression in pupils' learning. The sequence within a programme is merely the order in which content is introduced and activities are organised, whilst the idea of progression implies some advance in understanding, performance, appreciation or whatever. Progression therefore relates directly to what pupils learn, and its planning and evaluation require a clear conception of qualities and a good sense of direction. Progression is to be sought, not in some sequence of specific topics, but in the development of ideas, skills and sensibilities, for which the lesson content and activities should, at least in part, be designed.

The following simple propositions can form a starting point for a search for principles to guide a teacher who is seeking to strengthen the element of progression within a geography course.

1 Some ideas are inherently more difficult than others, some skills more demanding than others.
2 Many ideas can be approached at different levels of understanding.
3 An analysis of what is involved in particular learning tasks can help a teacher select appropriate methods.
4 Learning is facilitated by:
(a) building upon pupils' prior experience and knowledge;
(b) avoiding unnecessary difficulties or barriers.
5 Learning is facilitated when the experience provided and the tasks set by teachers are related to pupils' interests and capabilities, and the overall course programme is designed to take account of the way in which pupils mature.
6 Assessment of pupils' progress is necessary in order to adjust the programme to meet the requirements of individuals.

Within this set of propositions there are three basic components: the nature of what is to be learnt; the experience, interests and capabilities of the pupils; and the teaching/learning process (Fig. 28).

The objectives of a geography course are likely to be varied in character, and the ideas, skills and attitudes which are identified may be interrelated in complex ways. An analysis of what is involved in satisfying particular objectives and an appreciation of the difficulties presented by particular learning tasks would appear to be necessary for rational course

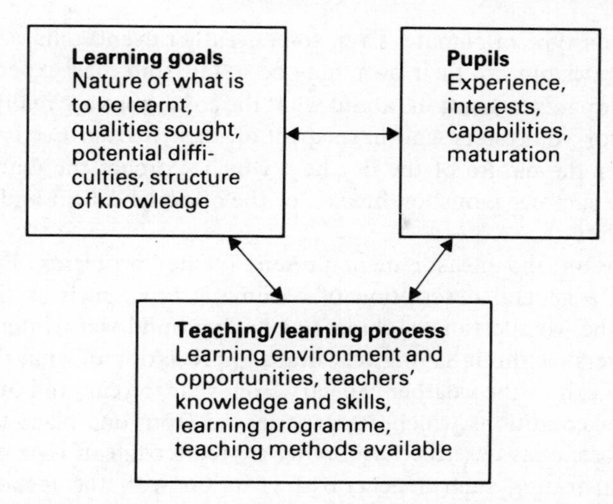

Fig. 28 *Elements which should influence the planning of progression within the curriculum*

planning. Pupils will vary in experience, interests and capabilities and will develop as a result of normal maturation as well as from a sound education. It is above all the teacher, through his planning, preparation and day-to-day management of the course, who must attempt to achieve a reasonable match between the capabilities of the pupils and the educational demands made upon them. Much depends on his knowledge and skills to use the opportunities available, to employ appropriate methods and to adjust the programme as necessary.

Learning difficulties

Returning to the first of the propositions, we can examine some familiar ideas which present learning difficulties and attempt to analyse why this should be so. The concept of 'climate' provides a good example. Whilst some of the problems which pupils face in studying climate may arise from inadequate teaching, others may be due to features inherent in the idea. Climate involves a wide variety of elements – air temperature, air pressure, wind, rain, snow, hail, cloud, sunshine, frost, fog, etc. – each of which vary in intensity or in occurrence. These elements combine and interact to produce weather conditions which in turn vary over space and fluctuate over time. If climate is conceived in terms of the pattern produced by weather over a period of years, it is more complex than weather insofar as it subsumes the latter. At the same time it is a more abstract idea, for whilst many weather conditions can be perceived as distinct events (e.g. a rain-shower, a thunderstorm, frost combined with fog on a winter morning), pupils must generalise from their experience to recognise a type of weather, and generalise yet further, and beyond direct experience, to appreciate a type of climate. Even some weather events, characteristic of other climates but not their own, may be so far from their experience that they develop misconceptions about what the conditions are really like. It is not easy for youngsters who have spent all their lives in Leeds to appreciate either the nature of the dry heat which scorches the Ganges Plain before the summer monsoon breaks, or the oppressive heat and humidity which follow.

Description and measurement present further problems. Pupils may memorise a general description of a climatic type, such as that which portrays the Mediterranean climate as having 'mild wet winters and hot dry summers', without having very much appreciation of what the climate is really like, how the weather can vary from year to year, and of the range of weather conditions which are experienced from one place to another within those areas described as having a Mediterranean type of climate. Whilst illustrating such a general description with the mean monthly temperature and rainfall figures of a few representative stations may in one sense provide greater precision, it does not help the pupils to imagine what the weather conditions are like, for these statistics by their nature conceal actual fluctuations. After all, they are designed to do just that. It is reasonable to assume that our use of a variety of scales to measure the properties of the atmosphere and our use of statistics to summarise climatic patterns present potential learning difficulties. When pupils are required to discriminate between the different climates of the world, they usually have to deal with about ten to twelve types and a classificatory scheme which is based on several criteria applied in a somewhat arbitrary way. It is hardly surprising that many appear to find this confusing.

Furthermore, the resultant distribution pattern is fairly complex, and, when mapped, boundary lines give the impression of sudden change, whereas in many areas there is a gradual transition from one type to another. Explanation of this complex pattern presents a further challenge, for a basic knowledge of physical principles is required to make much progress in understanding atmospheric processes, and abstract reasoning is required to grasp the nature of many of the relationships. To understand the distinctive seasonal pattern associated with the Mediterranean climate, one must appreciate the controls exerted by the westerlies, with their associated depression, in winter and the subtropical anticyclones, with their predominantly settled weather, in summer; to explain the distribution of this climatic type one must understand something about the general circulation of the atmosphere and the influence of latitude, the distribution of land and sea and of major relief features. This is fairly demanding, without even attempting to tackle questions about the nature of atmospheric stability and the causes of cyclogenesis. We need to consider very carefully what forms of 'explanation' are appropriate at different levels of understanding.

A second example of a familiar but difficult idea is that of 'conservation of natural resources', often introduced in relation to soil, water supply, forests and wildlife. The concept of a natural resource is far from straightforward, but that of conservation is more difficult. As a philosophy which is directed at the manner and timing of resource use, it has been subject to different interpretations and emphases (O'Riordan[2]; Simmons[3]). O'Riordan comments that it 'defies accurate translation into precise operational terms, and indeed suffers from being ambiguous in interpretation and shifting, inconsistent and sometimes contradictory in its perceived objectives'. Even the school pupil, who is not so concerned with matters of underlying philosophy, may find it difficult to distinguish between 'protection', 'preservation', 'wise use' and 'environmental management'. It is much easier to understand the nature of soil erosion, water shortages, forest depletion and the extinction of some form of wildlife, all of which can be illustrated in a simple and clear fashion, than to grasp the abstract idea of a carefully managed system within which there is stability. To understand how exploitation can co-exist with stability one must appreciate how the natural system functions and how it can be regulated. But many of these systems are very complex, as are the technological, economic and social factors which influence resource use and misuse. We can, of course, use conceptual models to help us make sense of the main features. Thus, in the field of water supply, the main flows, transfers and storages in a river basin can be described in terms of the basic hydrological

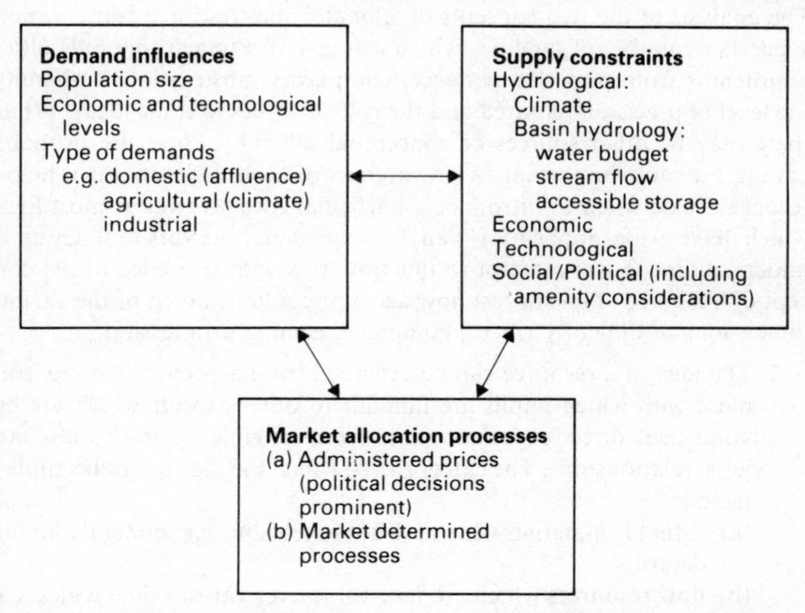

Fig. 29 *Schematic summary of the water supply resource system (following Manners)*

cycle, and an essential relationship within this system can be summarised in terms of the water balance:

Initial Storage + Precipitation = Evapotranspiration + Runoff + Final Storage

When we attempt to take account of the human factors involved in the use of water resources, we must include more elements and relationships (Fig. 29).

The apparent order and simplicity of the models are achieved at the cost of a fairly high level of generalisation and abstraction. Here pupils may face the same sort of learning difficulties as they do when studying climate. Resource conservation, however, presents a further problem, which has no parallel with climate, in that it is a value-laden concept. A notion such as 'stewardship', with its emphasis on man's responsibility for the environment and to future generations, has an obvious ethical dimension. But values and attitudes also influence our perception of resources, our evaluation of the benefits and costs of exploiting a given resource in a particular way, and our appraisal of any given environmental issue. This, as much as cognitive understanding, helps to explain the differences in interpretation and application of the idea. In this, 'resource conservation' is more akin to such ideas an 'environmental quality', 'welfare' and 'development' (Smith[4]).

Different levels of understanding

The analysis of the two concepts of 'climate' and 'resource conservation' suggests a number of qualities which can give rise to learning difficulties: remoteness from pupils' experience, complexity, abstraction, ambiguity, the level of precision required and the role of values in some ideas. Whilst there may be other sources of conceptual difficulty, these are probably among the most important. An awareness of such difficulties can help a teacher decide when to introduce a particular idea. However, most ideas which have wide application can be appreciated at different levels of understanding. I will attempt to illustrate this with reference to the concept of 'resource', and suggest how account could be taken of the various dimensions of difficulty when planning a learning progression.

1 The idea of a resource can be extended from aspects of the environment with which pupils are familiar to other aspects which are beyond their direct experience, and from tangible objects to less tangible relationships. The categories to which the idea might be applied include:
   - (a) valuable materials which are not renewable, e.g. mineral and fuel deposits;
   - (b) flow resources which are renewable over various time scales, e.g. soil, water, forests;
   - (c) localised natural features valued for specific purposes, e.g.

natural harbours, attractive beaches, sites for hydroelectric schemes;

(d) man-made resources, e.g. reservoirs, airfields, concert halls;

(e) human resources associated with the qualities of populations, e.g. skilled labour;

(f) less tangible components of the environment, e.g. climate, natural landscape, wilderness.

Over a period of time, pupils could be led to widen the concept to an increasing range of phenomena, and in so doing could be helped to reshape their original conception of what a resource is.

2 Selected case studies of the development of resources could be graded to allow for an increase in the complexity of the information supplied and understanding required about the conditions which influence demand, the factors which affect exploitation, the resource allocation procedures, the consequences of exploitation and the attempts at resource management and conservation. For example, having investigated some simple resource systems, pupils could be introduced to multi-purpose resources, such as forest areas which are valued for timber supply, watershed protection, grazing of animals, wild life habitats and recreation, in order to appreciate that resource management problems can be difficult and complex, especially when there are conflicting demands and values.

3 Some progression, which takes account of the demands of precision, might be planned in terms of the attention given to measuring the scale and quality of resources, and to measuring the costs and benefits of exploitation. For example, while in early studies general figures may be used to give an indication of the scale of mineral deposits in relation to the scale of mining, later studies could explore the difficulties of resource evaluation and 'the inherent limitations to the accuracy with which reserves can be quantified' (Manners[5]). Later studies could also give attention to the difficulties of predicting 'flow resources', and the problems of estimating the social costs and benefits of resource exploitation. A decision whether or not to have another large reservoir in a National Park is not simply a matter of technology and economics.

4 Abstraction may, however, be the most important dimension of difficulty of which account should be taken. In gradually extending the concept of resource to climate, landscape and wilderness, one is in any case moving towards greater abstraction. Furthermore, several important concepts which are closely associated with 'resource' are also fairly abstract: 'environmental perception', 'capacity', 'threshold', 'substitution' and, as emphasised above, the key ideas of 'conservation' and 'resource management'. To these may be added important spatial concepts such as 'accessibility'. As pupils become

capable of abstract thinking they can begin to understand that resources are culturally defined, and that appraisal of an environment and of the opportunities which it presents can change considerably over time, particularly as a result of technology and of changing cultural values. They can begin to understand how resource exploitation is influenced by spatial patterns of supply and demand, and how evaluation of the significance of a location can be changed by developments in technology, as well as by political and other factors. The specific knowledge acquired from case studies, combined with a more abstract analysis of relationships, can enable intellectually mature pupils to develop a clearer conception of the basic idea and recognise its importance within the conceptual structure of geography. The latter aim may be particularly appropriate at the level of the sixth form or in higher education.

Analysis of learning tasks

The analysis of an important concept, such as 'resource', can reveal:

1 the basic attributes of the idea;
2 the range of situations within which it is appropriate to introduce and utilise that idea;
3 the possible directions in which meaning can be extended and understanding can be developed; and
4 the difficulties which teachers and pupils can expect to encounter.

This could provide useful guidance for a teacher who is attempting to design a spiral curriculum, a programme in which key ideas and skills form 'threads which run through the curriculum in a cumulative and overarching fashion' (Taba[6]). Such threads, according to Taba, 'cannot be isolated into specific units but must be woven into the whole fabric of the curriculum and examined over and over again as an ascending apiral'. But the sort of analysis outlined with regard to 'resource' does not lead directly to a course programme. Rather, it reveals potential opportunities and constraints.

It is still necessary for the teacher, working with a particular group of pupils, to select from the range of opportunities revealed those which are most suitable for them at a given stage. Having determined the objectives, further analysis can clarify what it is that those pupils most understand or must be able to do if the objective is to be achieved. This in turn can help a teacher decide on appropriate content and methods, and on the way in which these can be organised in the form of a series of related learning activities. An analysis can indicate that a particular sequence of activities is reasonable, or even that one sequence is probably better than another, but, when educational objectives are complex, it is difficult to identify an ideal sequence. In the USA, empirical research has been stimulated by

Gagne's contention that an essential condition for the learning of an intellectual skill is the acquisition by the learner of a number of prerequisite skills and subskills, and that the whole structure can be represented by the learning hierarchy (Gagne[7]). The findings have not been altogether encouraging. Such are the difficulties in validating empirically any hypothesised hierarchy, that White[8] has concluded that 'it seems probably that only very narrowly defined abilities can be used in establishing a hierarchical relationship'. In fact, most of the research has been limited to fairly simple intellectual skills within mathematics and physical science, taught through a sequence of activities which are concentrated within a short period of time. Whilst such studies may contribute to the better design of a group of lessons planned around a specific objective, they would appear to be much less useful to planning overall structure of courses which extend over several years. The same may be true of structures based upon hierarchies of ideas. Several writers have attempted to identify conceptual hierarchies as a step towards planning a learning sequence (Tanck[9], Graves[10], McGattrick[11]), but most recognise that understanding can be reached through different routes. In many areas of knowledge, the complexity of potential links between ideas may defy a simple hierarchical arrangement. Geographers' use of the concept of a node in their analysis of spatial patterns and processes may serve as an example. As far as a geographer is concerned a node is a spatial concentration of activity, and the implication of using the term 'node' is that, for the moment, the dimensions of a particular object or place are ignored, whilst attention is focused on other attributes, notably those of relative location. The idea is especially useful

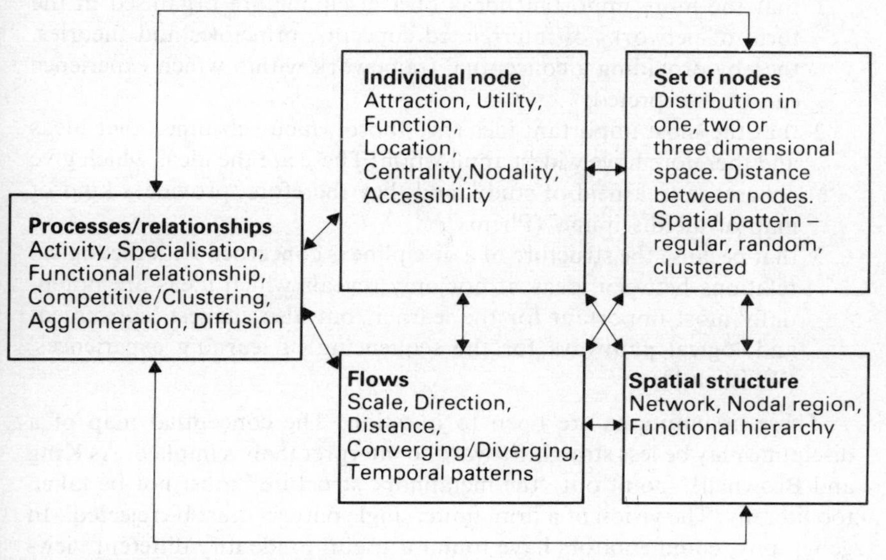

Fig. 30 *Ideas associated with the concept of a node*

in the study of settlements because human occupance is essentially focal in character (Philbrick[12]), with villages, towns and cities forming the central places serving their surrounding areas. In studies of spatial behaviour, nodes are defined by movement patterns – by flows of materials, people, money and information – and the importance of nodes may be measured by the strength of the flows and the size and nature of the areas which they serve. To have a reasonable understanding of how the idea is applied in geography, it is necessary to know how it is linked to other ideas. Figure 30 is an attempt to describe a framework within which such links can be explored.

It is difficult to conceive how such a wealth of ideas could be organised in the form of a single cognitive hierarchy, or even a small number of hierarchies. This is not to deny the value of identifying the important ideas with which a concept, such as node, is associated, and of clarifying the nature of the relationships between these ideas. Such an analysis, however, will not point directly to a learning structure based on some logical progression in understanding.

The structure of a discipline

The descriptions given above, of what is involved in developing a fuller understanding of the concepts of 'resource' and 'node', point to weaknesses in the suggestion, made by some curriculum theorists, that the key to the selection and sequencing of ideas may lie in the logical structure of a discipline. That suggestion appears to rest on several assumptions:

1 that the more important ideas of a discipline are organised in the form of networks of interrelated concepts, principles and theories, thereby providing a conceptual framework within which experience can be interpreted;
2 that the most important ideas are those which subsume other ideas and therefore have widest application. These are the ideas which give coherence to a field of study, and they therefore 'provide a kind of map of the discipline' (Phenix[13]);
3 that because the structure of a discipline is concerned with the logical relations between ideas, it not only reveals which ideas are potentially most important for the learner, but also suggest 'convenient and logical pathways for the sequencing of learning experiences' (Whitfield[14]).

These assumptions are open to question. The conceptual map of a discipline may be less straightforward to interpret than is implied. As King and Brownell[15] point out, 'the metaphor "structure" must not be taken too literally. The vision of a firm, finite, single pattern must be rejected'. In geography, commentators have found it useful to identify different viewpoints, each of which presents its own criteria for the selection and organisation of important concepts and principles (Pattison[16], Broek,[17] Haggett,[18] Harvey,[19] Taaffe[20]).

Furthermore, as in any vigorous discipline, new concepts are invented, new principles established and conflicting values are evident. To their traditional regional, ecological and spatial viewpoints, geographers may now wish to add the welfare approach to human geography (Smith[4]) or a more radical Marxist perspective. It all suggests a rather cluttered and untidy map. Even when specialists have a fair measure of agreement about the most important ideas and how they relate to one another, the conceptual structures which they share are the product of an advanced state of knowledge. 'The [conceptual] organisations employed by the expert are not necessarily the structures most useful in facilitating the learning of an individual at a particular developmental stage or at a particular level of sophistication in the subject matter' (Glaser and Resnick[21]); and a model which describes a set of relationships within a coherent body of knowledge does not necessarily prescribe the order in which learning should take place. The fact that many ideas can be approached at different levels of understanding greatly complicates the notion of a logical route through a 'map of knowledge'. In practice there are usually alternative routes and sometimes no clear indication which is the best. Furthermore, as the analysis of progression applied to the concept of resource illustrated, a number of return journeys may not be out of place, for the intellectual landscape can present a changing appearance to pupils who are growing in experience and maturity. Only gradually can they appreciate the abstract form of this landscape, and the value of their interpretation may be enhanced by memories accumulated from previous journeys. In a very real sense pupils construct their own maps of knowledge, and to be useful these maps must be related to their experience of the real world as well as to structures devised by scholars. Too many short-cuts may deprive them of the experience to cope with the task of interpreting other peoples' maps. Pupils, therefore, require routes which enable them to perceive some of the rich detail and at the same time acquire a sense of pattern and structure.

Pupil maturation

As pupils make their long, and often somewhat winding, educational journey, they gradually mature – physically, intellectually, socially and emotionally. Whilst the rate of maturation will be influenced by environmental conditions and learning experience, as well as inherited characteristics, it appears that the general pattern of development is similar in most individuals. Whether or not we accept Piaget's conclusion that children pass through a number of distinct stages, it is clear that as they grow older they develop different styles of thinking and problem solving, and they only gradually acquire the ability to appreciate some forms of reasoning. It is obviously important for teachers to be sensitive to their pupils'

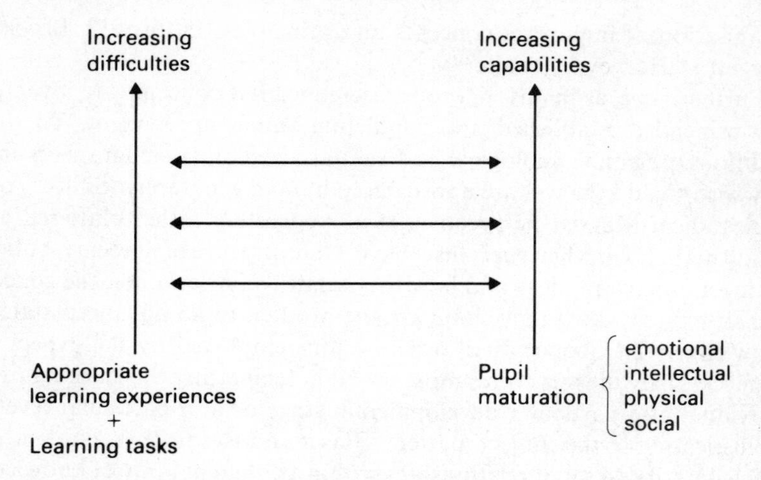

Fig. 31 *A matching model*

maturity and to match pupils' interests and capabilities with appropriate experiences and tasks.

Although comparatively little of the research into the development of children's thinking has been specifically geographical, there has been a growth of interest since the early 1960s (Ball,[22] Naish,[23] Saveland and Pannell[24]). Two examples will illustrate the insight which can be derived from empirical investigations of this type. The first is concerned with children's ability to read, interpret and draw maps. This is the area pertinent to geographers which has received most attention, and Catling[25] has recently reviewed the available research findings and outlined a development sequence for children's spatial understanding. He has provided a valuable indicator of what it is reasonable to expect pupils of primary school age to be able to do, and has suggested the sort of mapwork which could assist their learning. The second, which is of a more general nature, relates to the understanding and judgement of adolescents. Peel[26] has brought together the findings of a number of significant studies, including one which was geographical (Rhys[27]), which investigated the ability of older pupils to think in a formal way and to adopt a hypothetico-deductive approach to the solution of problems. He has shown the importance at this level of thinking of the ability to understand the nature of dynamic equilibrium; a difficult idea but one which becomes relevant in many fields of geography when the purpose of study moves from the description of recurrent patterns to an analysis of the processes that bring such patterns about. Part of the difficulty which pupils face when studying such concepts as 'climate' and 'resource conservation' is that a proper understanding of these depends on an appreciation of conditions which exhibit dynamic equilibrium. The more that climatic explanations are offered in terms of heat balances, lapse rates, atmospheric stability and instability, and the general circulation pattern, the greater the dependence on such understanding. It is therefore not surprising that even some sixth-form students find the climatology component of their Advance Level courses difficult.

The complexities of educational goals, of the structuring of knowledge within geography, and of the ways in which pupils develop; the variety of content and methods available; and the different opportunities and constraints with different schools, ensure that curriculum construction will remain as much an art as a science. Planning for progression in learning is an important part of that art. Whilst it can be assisted greatly by careful analysis of the sort of understanding and skills which teachers intend pupils to acquire, the task of designing a suitable programme must to some extent be a creative process, calling for imagination as well as analysis. Nevertheless, many teachers would probably welcome guidelines, not only in the form of a discussion of principles, but also in the form of some simple examples of conceptual frameworks which can be applied to geographical content.

The following is an attempt to outline a possible progression in one fairly popular topic, that of 'shopping patterns', a topic which pupils may return to several times in their geographical studies. Most children will, of course, have experience of shops and shopping from a very young age, and learning activities in the early years of schooling can help them to discriminate between different types of shops and to appreciate the nature of the transaction which takes place within these premises. Classroom simulations may help them to understand the roles of shop assistant and customer, as well as give them practice in a range of skills. The example, however, starts with the middle years of schooling when teachers are more likely to consider incorporating a specific geographical dimension.

A progression in the development of ideas on shopping patterns

Phase 1 (9–12 years)
In this phase, learning is likely to be more successful when based mainly on direct experience. This can include studies of:

(a) shopping behaviour – personal and family experience;
types of goods purchased at different centres;
frequency of visits;
distance travelled.

(b) shopping centres – small, local centres;
types of shops;
comparisons between centres.

Among the general ideas which might be introduced are:

(i) Some shops sell a great variety of goods (e.g. small general store, supermarket); other shops specialise in particular goods or services (e.g. butcher, greengrocer, toy shop, hairdresser).

(ii) Many families have fairly regular shopping habits (e.g. groceries may be obtained from the same shop or shopping centre at roughly the same time each week).

(iii) Some goods (e.g. fresh vegetables, bread and meat) are purchased more frequently than others (e.g. clothing, furniture).

(iv) Families tend to obtain those goods which they purchase frequently from shops near their home, but they may have to travel further, and are usually prepared to travel further, for goods which they only require to purchase occasionally.

(v) The grouping of a number of shops in one place is convenient both for customers, who can then easily visit several shops on one journey, and for traders, who may gain the custom of those travelling to the centre primarily to visit another shop.

vi The largest concentration of shops in a town is usually in the town centre. Here there are a greater variety of shops and many larger shops.

Phase 2 (12–14 years)

Pupils of this age are often capable of a more rigorous analysis of shopping behaviour and a fuller investigation and more precise description of types of shopping centres. They could, for example, map activity patterns for different members of a family, and they could examine the significance of multi-purpose trips. They are more capable of analysing the factors which influence customers' choice of shopping centres, identifying some of the opportunities and constraints which are taken into account by individuals. The analysis of a small centre could involve mapping its market area (shopping field), evaluating its location and evaluating its attractions as a centre. The analysis of a shopping field – a spatial pattern produced by a large number of individual movements – could be paralleled by studies of the catchment area of the school, the area from which a factory obtains its labour force, the area supporting a football club, and eventually the more complex idea of the 'sphere of influence' of a town. At this stage pupils might study a distribution pattern, analysing differences in the size and characteristics of a number of centres and measuring the distances between centres of similar size. How far measurement could usefully be employed would depend partly on the complexity of the local pattern. The identification of those types of shops and goods which are restricted to large centres would prepare pupils for the concept of 'order of goods', whilst the evidence of the spacing of centres and the patterns of shopping behaviour could form the basis for a consideration of 'range of goods' and 'threshold'. With pupils of this age, it is still important to base the analysis on concrete experience, but games and simulations may also help to focus pupils' attention on significant patterns and the decisions which help to produce these.

General ideas appropriate for this stage could include:

1 The decision where to shop is influenced by:
   (a) the sort of goods which the customer intends to buy;
   (b) the ease and cost of getting to alternative shopping centres, which depends mainly on:
      (i) the distribution of the centres;
      (ii) the distances which the customer must travel;
      (iii) the time and transport available;
   (c) the attractions of alternative shopping centres, e.g. what the customer knows or believes about the type and quality of shop, the price of goods, the choice of goods, parking facilities and other amenities.
2 Each shop has a minimum set of goods or services necessary to cover operating costs and provide an acceptable profit. Each shop and each shopping centre must therefore attract customers from an area large enough to support that trade. (A large chain store company may be prepared to operate a new shop at a loss for a period whilst attracting the necessary customers, but its choice of location will take account of the potential market in that area.)
3 Most small shopping centres consist mainly of shops selling 'convenience' goods (e.g. general grocer, greengrocer, butcher, chemist, baker, newsagent/confectioner/tobacconist, 'wool shop'). Such centres are numerous, with each drawing its customers mainly from its immediate locality.
4 Large shopping centres contain not only shops selling convenience goods, but also shops selling 'comparison' goods and 'specialist' goods. Those centres must attract customers from a larger area.
5 The combination of a few large centres, each serving a large area, and a much larger number of small centres serving small areas, which lie within the fields of the large centres, tends to produce a hierarchical pattern.

Phase 3 (14–16 years)

As pupils become capable of understanding more complex processes and relationships, it becomes reasonable to investigate

the nature of town centres, and

the processes which lead to changes in shopping patterns.

It is possible to look at town centres and changing patterns at a simple level with younger pupils, but at 14–16 it is easier to introduce appropriate conceptual models. Thus abler pupils can appreciate the processes influencing the CBD, as described in Fig. 32.

Although such an account is obviously incomplete, it is intellectually

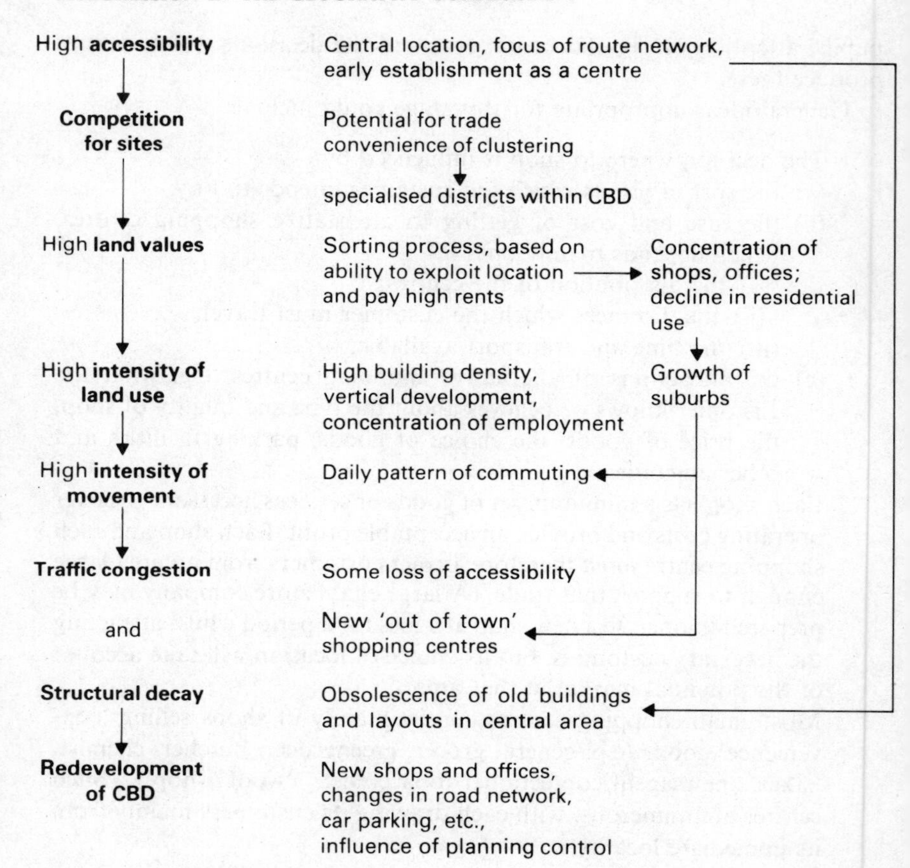

Fig. 32 *Processes influencing the CBD*

demanding insofar as it extends to features which cannot be observed directly and to processes which involve relationships between a number of variables. The thread of reasoning is dependent on a fairly abstract level of understanding. At the same time, it is likely that pupils of this age will be interested in the attitudes and goals of those who make the significant decisions affecting town centres, and they will develop their own views about the appropriateness of those decisions and the quality of the environment which results from them.

The considerable recent changes in the general charater and distribution of retailing and in patterns of shop behaviour likewise involve a variety of factors which are inter-related (Scott,[28] Kivell,[29] Parker,[30] Davies,[31], Newby and Shepherd[32]). Account may be taken of:

1 Changes in shopping habits and tastes. The higher level of car ownership, increasing the mobility and flexibility of shoppers, the growing popularity of freezers, the higher proportion of women in em-

ployment, and the greater proportion of income available for unessential goods all contribute to a tendency to shop less frequently and to make greater use of large centres which have a wide range of facilities. Customers may seek more pleasant environments in which to carry out their shopping.

2 Changes in the organisation and technology of retailing. There has been a marked trend towards self-service, an increased scale of operation, provision of better access for delivery of goods and provision of parking space for customers – features which are most evident in the growth of supermarkets and hypermarkets.
3 Shifts in the distribution of population and purchasing power. The effects of the decline of population in the inner areas of cities and the growth of suburbs have been amplified by socio-economic differences, resulting from the drift to the suburbs of the more prosperous families.
4 Changes in the accessibility of competing centres consequent upon changes in transport and road networks. In large cities accessible sites, located near important road junctions on or near the edge of the built-up area, can offer serious competition for older centres suffering from traffic congestion and inadequate parking.
5 The effects of planning decisions. In the United Kingdom, in contrast to North America and France, planning policies have restricted the development of large 'out-of-town' shopping centres and favoured the redevelopment of existing centres.

Here again there is scope for pupils to make qualitative judgements about the changes which have occurred and to consider future options.

Phase 4 (16–18 years)

Abler pupils who continue with the study of geography beyond the age of 16 should be capable of using more abstract ideas. In the study of shopping patterns this means that, given appropriate experience and guidance, they should be able to comprehend the concept of centrality and to make use of such abstract models as 'retail gravity model' and 'central place theory'. There is little point in attempting to deal with the assumptions and relationships which are embodied in such formulations until pupils are capable of propositional thought and are able to evaluate the models as aids to understanding. In other words, for the learning to be worthwhile, pupils must be able to appreciate the differences between the simplified elements in a model and the patterns and processes which can be observed in the real world, and at the same time understand what the model has to offer.

Some of the ideas which have been suggested as appropriate for the earlier phases would provide a firm foundation for the introduction of central place theory. Pupils would have studied the size and spacing of shopping centres, the spatial pattern of their market areas and patterns of shopping behaviour. They would have studied the factors which influence that behaviour, including distance, and they would have been introduced to the concepts of 'range of good' and 'threshold'. To appreciate Christaller's geometrical 'solution', however, pupils must understand the nature of the normative model and the significance of the assumption and principles which are built into this particular example; they must be able to grapple both with its internal logic and with the question of its empirical relevance (Harvey[19] Lewis[33]). The achievements of many students who have encountered central place theory in their GCE Advanced Level course appear to fall short of such requirements, as Tidswell shows elsewhere in this volume. Perhaps, although they have been given a description of the model, they have not adequately explored the underlying ideas. This is probably an even more serious risk when younger pupils are introduced to simplified, descriptive versions of 'Christaller', 'von Thunen', 'Burgess' or whatever. Whether an idea is appropriate to a particular phase of learning depends not only on the characteristics and form of the idea as presented but also on the style of thinking demanded from the pupils.

The general lines of development which have been suggested for the study of shopping patterns are summarised in Fig. 33. Obviously it is not the only sound sequence which could be devised, but it illustrates how the principles outlined earlier can be applied to a particular theme.

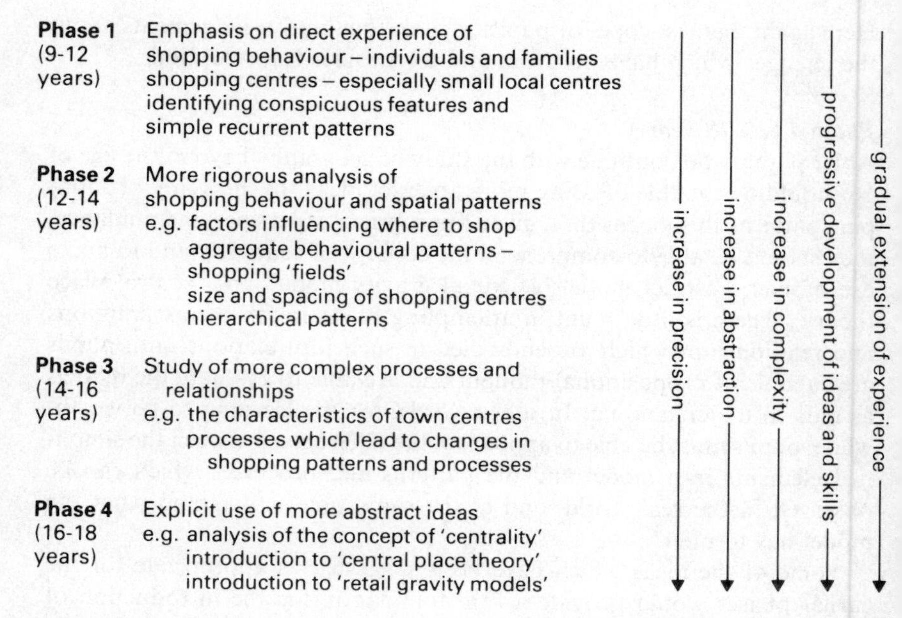

Fig. 33 *A progression in the development of ideas on shopping patterns*

Teachers would probably welcome rather more examples and some discussion of their relative merits. HMI have recently attempted to identify some of the main geographical ideas within major systematic themes such as farming, landforms, manufacturing industries and settlement which could provide suitable targets for secondary school pupils (HMI[34]). It would be useful to consider how such understanding could be built up over the period of compulsory schooling.

Evaluation

As pupils vary so greatly in their experience, motivation, capacity to learn and rate of development, and as we can never be sure that any particular teaching method will be successful, it is essential that the progress of individual pupils is adequately assessed and the findings of that assessment built into periodic evaluation of the teaching programme. However appropriate a general framework may appear to be, a proper sensitivity to pupils' responses and to circumstances which cannot be foreseen will almost inevitably call for some adjustments to a curricular plan. As progression in learning is concerned mainly with pupils' development of understanding and skills, geography teachers need to develop appropriate techniques of assessment. Without appropriate assessment of pupils' learning we may be fooling ourselves that progression is being achieved.

Note:
The views expressed in the chapter are those of the writer and do not necessarily reflect the views of the Department of Education and Science.

References

1 BLYTH, A. et al. (1976) *Curriculum Planning, History, Geography and Social Science*. Collins
2 O'RIORDAN, T. (1971) *Perspectives on Resource Management*. Pion
3 SIMMINS, I. G. (1973) 'Conservation' in Dawson, J. A. and Doorkamp, J. C. (ed.) *Evaluating the Human Environment*. Arnold
4 SMITH, D. (1977) *Human Geography. A Welfare Approach*. Arnold
5 MANNERS, G. (1969) 'New resource evaluations', in Cooke, R. U. and Johnson, J. H., *Trends in Geography*. Pergamon
6 TABA, H. (1962) *Curriculum Development. Theory and Practice*. Harcourt Brace
7 GAGNE, R. M. (1977) *The Conditions of Learning*, Holt, Rinehardt and Winston, 3rd edn. (The first edition was published in 1965.)
8 WHITE, T. R. (1973) 'Research into learning hierarchies', *Review of Educational Research* **43** (3)
9 TANCK, M. L. (1969) 'Teaching concepts, generalisation and constructs' in *39th Yearbook of the National Council for the Social Studies*
10 GRAVES, N. J. (1975) *Geography in Education*. Heinemann

11 McGATTRICK, B. (1976) 'Learning theory and assessment in geography teaching', in Dunlop, S. (ed.) *Place and People*. Heinemann Educational Books
12 PHILBRICK, A. K. (1963) *This Human World*. Wiley
13 PHENIX, P. H. (1964) *Realms of Meaning*. McGraw-Hill
14 WHITFIELD, R. C. (1971) *Disciplines of the Curriculum*. McGraw-Hill
15 KING, A. R. and BROWNELL, J. A. (1966) *The Curriculum and the Disciplines of Knowledge*. Wiley
16 PATTISON, W. D. (1964) 'The four traditions of geography', *Journal of Geography* **63**(5)
17 BROEK, J. (1965) *Geography. Its Scope and Purpose*. Columbus
18 HAGGETT, P. (1965) *Locational Analysis in Human Geography*. Arnold
19 HARVEY, D. (1969, 1972) *Explanation in Geography*. Arnold. 'The role of theory' in Graves, N. J. (ed.) *New Movements in the Study and Teaching of Geography*. Temple Smith 1972
20 TAAFE, E. J. (1974) 'The spatial view in context', *AAAG* **64**(1)
21 GLASER, R. & RESNICK, L. B. (1972) 'Instructional psychology', *Annual Review of Psychology* **23**
22 BALL, J. M. (1969) 'A bibliography for geographic education', Geography Curriculum Project, University of Georgia. Athens, Georgia
23 NAISH, M. C. (1972) 'Some aspects of the study and teaching of geography in Britain. A review of recent British research', *Teaching Geography Occasional Paper No. 18*. The Geographical Association
24 SAVELAND, R. N. and PANNELL, C. W. (1978) 'Some aspects of the study and teaching of geography in the United States, a review of current research 1965–1975', *Teaching Geography Occasional Paper No. 30*. The Geography Association
25 CATLING, S. J. (1979) 'Maps and cognitive maps: the young child's perception', *Geography* **64**(4)
26 PEEL, E. A. (1971) *The Nature of Adolescent Judgment*. Staples Press
27 RHYS, W. T. (1972) 'The development of logical thinking' in Graves, N. J. (ed.) *New Movements in the Study and Teaching of Geography*. Temple Smith
28 SCOTT, P. (1970) *Geography and Retailing*. Hutchinson
29 KIVELL, P. T. (1972) 'Retailing in non-central locations', *Institute of British Geographers Occasional Publication* (1)
30 PARKER, A. J. (1975) 'Hypermarkets: the changing patterns of retailing', *Geography*, **60** (2)
31 DAVIES, R. L. (1976, 1978), *Marketing Geography with Special Reference to Retailing*. Methuen. 'Shopping as a family expedition', *Geographical Magazine*, February
32 NEWBY, P. T. and SHEPHERD, I. D. H. (1979) 'Brent Cross: a milestone in retail development', *Geography*, **64** (2)
33 LEWIS, R. (1977) 'Central Place Analysis', Unit 10 of the Open Univer-

sity Course. *Fundamentals of Human Geography*. The Open University Press

34 HMI (1978) 'The teaching of ideas in geography. Some suggestions for the middle and secondary years of education', *HMI Matters for Discussion* (5). HMSO

10 Educational computing and geography

David Walker University of Loughborough

The 1970s saw the beginning of educational computing in geography. It was a period of experimentation and innovation, though restricted to a few schools having access to the rare and expensive computers of the time, and restricted also to a limited range of application. In the 1980s the revolutionary micro-processor, which is so influencing industry and commerce, will open up new opportunities, present new challenges and above all make computers widely available for use in geography teaching.

Traditional computers cost more than a light aircraft, and needed considerable skill to operate. Although they were very powerful in the amount of work that they could do in a short time, they were essentially vast data processing machines and not well suited to educational applications.

The micro-computer of the 1980s now costs about the price of a motorbike and prices are continuing to fall. Micro-computers are so necessary to many aspects of commercial and industrial life that it is important that pupils should be familiar with their power and their limitations. The substantial government funding for the development of educational computing will make sure that there is no shortage of materials.

For the geography teacher the initial attraction of the computer was that it could undertake the routine and the boring, which so often formed a barrier to effective styles of learning. The early excitement of getting into real problem solving and using actual field work data was sadly diminished by the tedious hours of doing the calculations by hand. There were genuinely some good reasons for resisting the quantitative revolution. But although the use of the computer for statistical work is the obvious first application it is not the only one, and in the long run may not be the most important.

For the geography teacher and the student, there are three ways in which the computer will have an impact on the teaching of the subject. Computers will become a part of educational technology, used both to demonstrate and as a basis for exercises to reinforce learning in much the same way that laboratory apparatus can be used for demonstration and for experiment in the natural science subjects. Second, computer use will become an essential skill for geographers, comparable to drawing maps or consulting reference books (both of which activities may become computerised). Geography teachers will have to learn and teach these skills and be as competent advising a pupil on the use of the computer as they are on the use of the mapping pen and the reference library. Finally, on a rather longer time scale, the computer is going to change the geography which we teach. Silicon Valley in California is growing at a pace which is reminiscent of the growth of the older industrial areas of Britain. Yet perhaps the most significant change is going to be caused by the reduction in labour requirements in industrial and commercial activities which have traditionally been major employers. The micro-processor is at the heart of the new telephone systems that will make electronic communication fast, reliable and cheap, reducing the need for people to work together. Economies of scale will become less important and the pressures leading to urbanisation will decline. New industry will move to areas where there is a supply of well-educated labour, or which has a good enough image to attract the high quality labour that will be needed to design and maintain computer-controlled systems.

Steam power heralded the first Industrial Revolution, and computers are the heralds of a new revolution which is going to have equally important economic and social consequences. The educational system cannot afford to ignore this revolution as it ignored the first.

Yet most of the public and many teachers have ambivalent attitudes to the computer, only partly Luddite in origin. It really is quite difficult to conceive how the computer can be of use in a teaching situation. Its purpose and operation is not clear from its appearance and design (a slide projector proclaims its purpose by its shape). Even those who have been working with computers for years are able to think of new applications and wonder why no one has ever thought of them before. The difference from all other machines is that the actual operation of the machine is controlled by a series of instructions, thought up in the head, scribbled down on paper, then typed in and stored in the machine. It takes some time to grasp that designing programs to control computers is not as easy as working a slide projector, but considerably easier than most people imagine.

Most of the information about new developments in computing that is reported in the press and on television relates to developments in electronic hardware. These have been occurring with startling rapidity over the last ten years and have opened up great new opportunities. However, the future is going to be influenced considerably more by the type of programs that are going to be developed, by the imagination of the programmers and the extent to which teachers will recognise the potential and make use of computers as a regular resource for teaching.

One of the major limitations of the old style of computer was that it was large and remote. There were two ways of getting it to work for you. One was to send instructions and data by post, carrier pigeon or runner and then wait for days, or even weeks, for the result to come back. This method was pursued by only the most dedicated and had very limited potential. The second method was to connect up the computer by telephone and remote typewriter, much faster but still slow and limited to typing messages in and out.

The microcomputer of the 1980s is not only much cheaper, it is actually much better for teaching purposes, and the indications are that it will continue to get better and cheaper during the coming decade. Accessibility was the big problem with the old generation of computers. The new machines can be carried to where they are needed (and some can literally be carried in a large pocket and used from rechargeable batteries with obvious implications for field work). As the price falls to that of a good bicycle it is going to be reasonable to have several in each school or even one in each department. Currently micro-computers in school are at the head of a long queue of eager pupils and strict time-tabling has to be enforced to let everyone have a turn. This is a great contrast to the fate of most educational resources which languish most of their lives in cupboards. So long as they are not being used for trivial purposes there is a strong case for saying the more computers the better, until the point is reached where there are enough to satisfy demand.

Another big problem of the old computers was that the results of calculations were displayed fairly slowly and only in the form of typewriting. The micro-computer can communicate directly with a television screen and can produce its messages so rapidly that it has to stop at the end of each screenful to give the user a chance to catch up. Because the screen is close to the machine the contents of each printing position is usually held in the internal memory in the computer, so it is possible to alter any one character at a time. This means that not only can tables and lists be altered to show the effects of changes in selection from the data, but charts and graphs can be built up and dynamically altered to show the working of processes. With enough memory it is possible to control each of the 7×8 dots that are used to make up each character to produce high resolution graphics with full colour so that maps, cross sections and diagrams can be calculated and displayed.

The limitation of the old style computer which would only take typed instructions can also be overcome by the micro-computer. Relatively cheap digitisers are available which will allow information about a map to be fed into the computer as a pen is traced round the outline. The digitising board can be used to hold a list of commands to make the program run and by touching the right command with the pen the computer will respond. The ability to type is always a useful skill, but it should not be a barrier to effective use of the computer.

Although the screen is good for demonstration and for some investigatory work the transitory nature of the image does have limitations, just as television has limitations in comparison with a book. If a record is needed the computer can be connected to a printer. Printers have come down in price dramatically and now cost less than a micro-computer (it used to be the other way round). Because they print using a matrix of pins it is possible to reproduce the images and diagrams from the screen (though not in colour).

One of the advantages of the big computer is that it can hold very large lists of information on external magnetic storage; there are economies of scale here because the mechanism to drive the record/playback machine is high precision and expensive; increasing the amount of information that can be stored is proportionately cheaper. The user of the micro has several interesting ways of overcoming this problem. The first is to make use of someone else's big computer to get hold of the information that is to be analysed on the micro-computer. In the future one way will be to use the Post Office Prestel system; it will even be possible to broadcast the data, to record these and then feed them into the computer. The other approach is to use a cheaper method for data storage, and preferably one that has had its development costs covered by sales for another purpose. Video cassette recorders can be adapted for computer storage and can hold vast amounts of information, but are likely to be used for a reserve store since it is not easy to locate one particular piece of information on demand.

The other development likely to be of great potential interest is the rival of the video cassette recorder, the video disc. The information needed to create an hour-long television programme in colour can be stored as small indentations on the surface of the vinyl pressing and read by analysing the reflections from a small laser beam. One hour's TV consists of the equivalent of 90 000 still pictures and this amount of information can be stored on each side of the disc. The scope of storing sets of programs and large sets of data, even the equivalent of pages from a book, is very dramatic. The development costs are paid for by the applications that require to flash through the information at twenty-five frames a second to get a moving picture. Forget the animation and those 90 000 frames a side are there for information storage and rapid retrieval, all on a disc which will cost a trifling amount to produce and distribute. The video disc can also bridge the gap between computing and visual aids since many of these frames could be used to hold still pictures which could be searched for and retrieved within seconds.

The overlap between educational technology and entertainment technology is going to produce other benefits. We are now seeing the first of the home TV games that have a keyboard and which can be programmed. The TV game is a simple microcomputer and the barrier between it and the microcomputer as seen in school is bound to blur. By the end of the 1980s as many children will be able to use the home computer/entertainment centre for their studies as now are able to watch television; and television has already shown that it has potential to change the whole set of knowledge and attitudes of a generation of children.

It is clear from all this that the computing of the 1980s will not be merely the mimicking of the work on the old generation of computers on new cheap machines. There are vast new opportunities for the geography teacher to make use of the most powerful of intellectual tools; and just as geography teachers have always known that it is part of their job to teach general skills of literacy, problem solving, and numeracy, so they will be involved in teaching the general skills of effective computer use (even making sure that students can use the word-processor effectively to organise punctuation and to correct spelling).

So far the discussion has been about the developments that have taken place in the technical aspects of the computer, none of which is going to make much impact unless programs are designed that will make this general-purpose tool relevant to the need of the geography teacher. The initial stage must be to make sure that everything that has been possible on the old computer is also possible on the micro-computer – the established collection of computer programs that do the hard work in statistical calculation, that do the book-keeping and that keep the rules in the more sophisticated games and the programs that work out the predictions of the locational models as pupils change the values of the variables. All of these can be used equally well on micro-computers. Most of them are worth modifying so that they can take advantage of the graphical output and instant response of the computer; the provision of graphs and illustrations is now relatively simple and could be added in to existing programs.

Capitalising on the power of the computer to store and retrieve information, to perform calculations and to sort through data and to present the results in pictorial and diagrammatic form – the potential applications can now go considerably beyond those that have been developed so far. At the start of any new period of development it is important to stand back and to ask why we want it and what the pupils will gain from it, to think carefully about the purpose of geography teaching and then to see how the computer can help to serve that end.

At the simplest level the role of the subject is to enable students to understand how the world has developed, in terms of landform, climate and vegetation, landscape, land use and human economic activity; to understand the processes that have operated and are operating and to be able to analyse the problems that are faced by mankind, and have enough sound knowledge to be able to understand the issues and to support policies for the effective management of the environmental and spatial resources. It is essential that the geography teacher keeps a firm eye on the contribution that the computer can make to the education of the pupil and make use of the computer when it will serve that end. Some examples will illustrate the type of program that will take advantage of the new technology and assist in the teaching and learning of useful geography.

1 Flooding is a serious problem for some areas of the country and very large sums are spent on flood prevention; some of the flood damage has been caused by land drainage and by building on land that used to have a temporary holding capacity. A computer model can show the pattern of streams and the rates of flow up to and including flood levels as a storm passes across the region. The effects of the rainstorm will vary, depending on the amount of water-holding capacity available in the soil and this can be varied to see the typical effects. Embankments and new cuts can be added to the system and the effects of these can be shown.
2 Large files of information derived from grid squares census data can be used to produce a series of thematic maps of the local area to demonstrate distribution patterns of census variables. Hypotheses about likely relationships and concentrations can be investigated using statistical procedures and graphing the results, as well as displaying the results back on the map to show the areas where the relationship holds and those where it does not so that a problem-solving approach by hypothesis testing can be followed through.
3 One of the problems faced by teachers has been that it has been very difficult to do much with some of the theories that are now considered to be part of the main stream of the subject; they are difficult to explain and difficult to test. They often cause real problems because pupils either believe that they are explanations of the real world like the theory of gravitation, or that they are totally fanciful. The computer can take real world data about transport costs, distances, numbers of people in settlements, etc. and make predictions for special cases based on theoretical approaches which can then be compared with the real world pattern. Pupils can then see just how close, or remote, the explanation provided by the theory is. They can quickly get into the most exciting part of theoretical geography, that of trying to understand why there are exceptions and to try to understand the complex interlinking of generalisations and specific events that has made our landscape. The conceptual grasp of the real world patterns and of the patterns produced by the theoretical model are of course much simpler when they can be shown in map form on the TV screen.
4 The processes operating with air masses to produce upcurrents,

clouds and rainfall can be difficult to teach and illustrate. Graphic displays can show the effects of changing the atmospheric lapse rate and the dew point, with some combination producing stable conditions and others producing cumulus clouds and precipitation. The model can be stopped and temperatures displayed at different times during the growth of the cloud and the lapse rate can be shown diagrammatically.

During the 1980s the computer will become an increasingly important part of everyday life, both at work and for entertainment. In geography there is great potential, increasing with the capabilities of the machines. All it needs is imagination and a system for developing and publishing computer materials. The latter exists and the former will grow as our understanding of the potential grows.

---

11 Towards a new generation of teacher-technologists

Michael G. Day The Haberdashers' Aske's School for Boys, Elstree

The aim of this paper is to consider the impact of the recent and rapid growth of educational technology upon the young geography teacher. Educational technology can be defined as the whole range of teaching aids, from duplicated worksheets to lithographed work units, from slide projectors to video cassette recorders, and from hand calculators to desk-top minicomputers.

The growth of educational technology over the past 15 years

In 1965, UNESCO published a *Source Book for Geography Teaching*[1] which readily acknowledged that visual aids and other educational technology had an important contribution to make to pupils' learning ability. Accordingly, they printed a list of what they considered to be necessary equipment for any geography department. This list has been reproduced in Table 17. Note that at the time of publication the authors thought it very unlikely that any more than just a few schools would have much beyond the 'minimum equipment' part of the list.

However, if we now compare this list with similar inventories in Table 17 of the educational technology owned or used by two secondary school geography departments in the South East, we can see that the contrast in the range, sophistication and type of equipment used in the classroom is quite striking. Clearly, educational technology has been a growth industry

Table 17 *A comparison of education technology inventories*

1965 *A list of required teaching materials for a typical geography department – from the UNESCO publication,* A Source Book for Geography Teaching		*1980* *An Inner London Comprehensive School – 1 000 pupils. (Geography taught as part of Social Studies Department)*		*1980* *Direct grant Grammar school, going Independent – 1 300 pupils, Greater London*	
Minimum equipment	chalkboards notebooks textbooks atlases globe wall maps meteorological and cartographic instruments specimen collection	Department equipment	whiteboards notebooks/textbooks atlases/globes/ wallmaps map chests one overhead projector for each classroom three slide projectors four radios four tape recorders and tape library black and white plus colour video cassette recorder colour television video cassette library three 16 mm film projectors six hand calculators	Department equipment	revolving chalkboards notebooks/textbooks atlases/globes/ wall maps map chests Geography library – 2 000 books, plus periodicals and press cutting service one overhead projector for each classroom four slide projectors with remote-control facility in each room a slide library, partially completed from cut-up film strips radio-cassette recorder for classroom use

			fieldwork instruments tracing table		tape library of commercial and school recorded material on audio and visual cassettes class set of hand calculators plus two desk top calculators 8 mm film projector weather station fieldwork instruments tracing tables manual photographic enlar- ger and reducer
Optimum equipment	epidiascope slide projector spirit duplicator cine film projector television map collection geography library	Supporting school equipment	video cassette recorder colour television 8 mm cine camera two photocopiers developing facilities school library with geography section computer with one master visual display unit and one terminal. one desk-top microcomputer	Supporting school equipment	colour video cassette recorder and television video rover camera dark room facilities one dry photocopier, plus one double side photocopier offset litho printer 16 mm film projector powerful computer with 1 master plus eight visual display units lecture theatre, fully equipped

over the past 15 years, and basic resource management and retrieval problems have now become part of the geography teachers daily routine.

The growth of educational technology has also directly affected classroom life. Consider a typical lesson c. 1965. The example illustrated here is a synposis of a 'model' lesson plan published in the UNESCO handbook.[2]

Aims:	1 To discover the population distribution of Malaya.
	2 To consider the distribution of rubber cultivation in Malaya
Equipment:	wall map of Asia
	chalkboard
	atlases
	textbooks
Method:	1 Oral question/answer exercise – wall map and atlases: Where is Malaysia? Where do most people live? Why?
	2 Dictated notes on Malaysia's population
	3 Oral question/answer exercise – chalkboard map and textbooks: Where is rubber grown? What is the relationship with climatic statistics? Why is rubber important to Malaysia?
	4 Written notes from headings on board
	5 Concluding link – other aspects of Malaysian life will be considered next week

How would the same topic be represented today? Inevitably, the teacher would incorporate a much wider range of educational technology into the lesson plan, perhaps selected from the following list of possibilities:

1 A BBC radio-derived audio tape and slide set on rubber plantations.
2 A home-produced audio tape.
3 A BBC video tape on Malaysia or rubber plantations.
4 A commercially available 16 mm film on Malaysia or rubber plantations.
5 A simulation game on plantation agriculture, perhaps played on a desk top mini-computer.
6 A set of pre-prepared overhead projector maps, diagrams and notes.
7 A pre-prepared photocopied worksheet, with durable master.

From the point of view of the pupil, this approach will certainly have been a much more stimulating lesson. He would see and hear the real world brought into the classroom, be presented with a wide range of activities during the lesson and be provided with carefully prepared and considered resource material, rather than hurriedly drawn blackboard maps and notes.

The implications for the teacher, however, are somewhat different. The teacher's skill in assembling and co-ordinating such a high technology lesson could be severely taxed, especially since such organisation needs to take place well in advance, and his success will often depend upon the efficiency of his own department in handling and storing equipment. On the other hand, more pre-prepared material, most of which can be stored for later use, should free him from extended lesson preparation for later lessons, and allow him to concentrate upon the needs of the individual pupils concerned. In other words, the teacher becomes more of a manager and consultant, rather than pedagogue.

It is clear that for a young teacher entering the profession today, a clear understanding of the potential of educational technology, and most effective way of using it in a given situation, is of paramount importance. Does current training, therefore, reflect the importance of technology?

Are student teachers adequately prepared in the use of educational technology?

My own experience is perhaps relevant here. I gained my Post Graduate Certificate in Education in 1976, having graduated a year earlier with an Honours degree in Geography. The university department of education that I attended was very much aware of the importance of educational technology. From a series of core lectures, I estimate that no less than 20 per cent were directly concerned with this topic. Further to this, a specific course on the uses, misuses and application of educational technology was run, although this was unfortunately optional due to the limited facilities available. Perhaps most important of all, an enthusiastic tutor ensured constant dialogue between the students throughout the year, which often centred around the practical application of technology in the classroom. I certainly felt well prepared when I began teaching practice, and subsequently went on to experiment with technology, and incorporated it where I felt it was relevant into my teaching.

This happy experience during training seems to be paralleled by most other young geography teachers entering the profession. But a sense of bewilderment often sets in when the student embarks upon teaching practice, or takes up full-time employment.

The following extract is from the transcript of a discussion with a young geography student during teaching practice. Just how typical are these sort of comments?

Geography student, age 22, studying for Post Graduate Certificate in Education
Mixed comprehensive school, 1 100 pupils – Hertfordshire
February 1980.

Q – So how well prepared do you think that you were for using educational technology before you started teaching practice?

A – I thought that I was well prepared ... I mean, because we had done a few simple example lessons, and discussed them in tutorials, but we hadn't really looked at pitfalls.

Q – What do you mean by pitfalls?

A – Well, take this school for example. If I want to use the projector, I have to find the woman who has the key to the audio-visual room. Last week, she was away, and I could not find a duplicate key, and that ruined my lessons for the day!

Q – Which aid do you use most often?

A – The overhead projector.

Q – Have you used the slide projector since you have been here?

A – Only when the tutor came round! No ... honestly, really, more than that – perhaps once or twice. But as I said, it is such a problem to get hold of the equipment, and to then cart it half way around the building, I can't be bothered. There isn't that much time when you are teaching, is there? I know I should use it more often.

Q – Do you find the same problem with other pieces of equipment?

A – Yes, as I was telling you before we started, my department has a large stock of films on video cassette, but you try getting hold of the machine to show them to the children!

Q – What do you feel about that?

A – Well, it's bad isn't it?

Q – What other technology have you used during teaching practice?

A – I've used the tape recorder once or twice ... but that's it, really. The tapes weren't all that good.

Q – Could you have made your own tapes, instead?

A – Yes, I've already done that once. I used my own recorder too!

Q – Do you think that you have used educational technology enough since you have been on teaching practice?

A – No – I would have like to have used it a lot more if I could.

Q – Have you felt that using technology extends your preparation time?

A – Yes – a great deal. But on the other hand, I think once I've got myself sorted out, and know where everything is, I don't think it would be that important any longer. Besides, they're not really all that organised here. It's getting hold of the stuff in the school that's the problem.

During this interview, it had become clear to me that here was a student who had been adequately trained, and was certainly aware of the potential of educational technology. However, frustration had set in when she tried to apply this initial enthusiasm during teaching practice in a school which did not share her interest in the use of technology. Is this where the stumbling block lies? It seems to me that this problem could be the link between the initial enthusiasm of the student, and an increasing apathy towards educational technology some years later on.

What happens to the initial enthusiasm of the young teacher for educational technology?

To begin to answer this question, consider the information shown in Table 18 which shows how often some typical items of technology are used in an Inner London Comprehensive school.

The data was gathered from a questionnaire survey of fifty teachers in an Inner London Comprehensive school of 1 000 pupils. Each entry

Table 18 *The intensity of use of a range of teaching technologies*

	Yesterday	*Within past 3 days*	*Within past 7 days*	*Within past month*	*Occasionally*	*Never*
Overhead projector	2 (3)	3 (1)	18	19	3	5
Spirit duplicator plus photocopier	18 (2)	10 (2)	20	0	2	0
Slide projector	0	0 (3)	8	15 (1)	11	16
Audio tape recorder	1	0	3	2 (1)	3 (3)	41
Video tape recorder and television	12	14	18 (2)	0 (1)	3 (1)	3
16 mm film	3	11	21 (2)	3 (1)	2 (1)	10
Computer	1	0	1	0	0	48 (4)

indicates the number of replies for that category, in answer to the question 'When did you last use the following items of equipment during teaching?' Brackets ( ) indicate the individual result for the geography department in this school, which comprises three members plus one part time.

The trends revealed by Table 18 would seem to indicate a certain degree of under-utilisation of the available technology, and it is interesting to examine the distribution of these results for each item in some detail.

The overhead projector. This was, surprisingly, not as well used as it might be, especially considering the basic nature of the equipment in any modern classroom. The majority of teachers had only used it occasionally during the past month, rather than it being in constant use. I felt that this was because the teachers tended to over-emphasise the advantages of duplicated worksheets when teaching mixed ability sets. The result seemed to be rather a bland diet for the pupil, with little variation to stimulate the imagination.

The duplicator and photocopier. Not surprisingly, considering the above comments, these items were very heavily used indeed.

The slide projector and audio tape recorder. This was a very disappointing result, with a very high proportion of teachers admitting to little regard for these items of equipment, especially the use of audio tapes. However, one very important consideration was often mentioned - the shear logistical problem of actually getting hold of the equipment, and then transporting it some distance to the classroom concerned. This is not a trivial consideration where bulky equipment is concerned, but the small modern audio tape player would surely solve this problem quite well.

Video tapes and 16 mm films. A high degree of use in this section is perhaps a reflection of the type of child being taught. However, it was also clear that constant use meant that the pool of material was quickly emptied, and many of the films were old, and inappropriate.

The computer. The school possessed two computers - one an early but powerful machine with printed output only, but the other a modern desktop computer with a visual display, and a simple keyboard input. It was clear that few pupils in the school used either of these machines, and many were not even aware of their existence. A number of teachers considered the equipment inappropriate for their type of school. This was perhaps rather unimaginative, and a consequent waste of a valuable resource.

The results quoted here are from a limited survey in one particular school. But would similar surveys elsewhere produce similar results? I believe so. If this is true, then it is a rather depressing situation. It would seem that operational problems and a general lack of imaginative understanding has lead to under-use or even misuse of a great deal of educational technology. And how long does it take for the young teacher to get caught up in the general syndrome?

Consider the following extract from the transcript of a taped interview with a young geography teacher, working in the same school from which the data in Table 18 was obtained.

Geography teacher, degree, qualified in 1976.
Mixed comprehensive school, 1 000 pupils, Inner London
March 1980.

Q – Which piece of educational technology do you use most often?

A – The overhead projector. (Interesting – note the survey results in Table 18.)

Q – Why?

A – Well, it's more convenient than the blackboard. The boards in this school are very pitted and in a bad condition, so it's much clearer for the children.

Q – What do you use the overhead projector for?

A – Mainly notes. I also like to show the children video tapes. We have a lot of discipline problems in this school, and videos are attractive. In any case, they help to bring the outside world into the classroom.

Q – What other aids do you use?

A – Films. ILEA has a large library that we can use.

Q – Have you ever used a tape-recorder in the classroom?

A – No, not really – I did on teaching practice, but it is difficult to get hold of the machine here.

Q – Do you ever use any of the commercially available teaching tapes, such as those produced by the BBC?

A – I think we have a few in the department.

Q – Do you use your own slides, or do you use commercially available slide sets?

A – Well, to be honest, showing slides is a joke in this school. If you can find a projector that works, someone else is borrowing it, and the Media Resources Officer insists on setting up any equipment himself. I used to use them a lot when I first came here, but I don't now.

To me this backs up the initial hypothesis – that young teachers are prone to lose their initial enthusiasm for technology fairly quickly when faced with frustrating outside factors in their first school. There is no doubt that when the geography department has adequate resources and is competently organised, the teacher can and does use technology in a most imaginative way. But without this logistic support, the use of technology becomes frustrated, and the aggravation of inadequate equipment and unprofessional organisation takes its toll.

But how does all this affect the pupil? Does the pupil appreciate the increased use of technology in his or her lesson – indeed, is it expected? This is also clearly a key issue, since if we cannot measure a distinct response from the pupil, then educational technology can no longer be considered cost-effective.

Is educational technology cost-effective?

A difficult question such as this demands a closely monitored study, far beyond the scope of this paper. Nonetheless, consider the following:

Transcript of interview
Pupil aged 13, Hertfordshire comprehensive school, taught by young geography teacher

Q – Do you like geography?
A – Yes.
Q – Why?
A – I enjoy it, it's interesting.
Q – Do you like watching some of the video films you see?
A – Yes.
Q – Why do you enjoy that?
A – . . . Well, it's better than just listening to someone just talking all the time.
Q – What else do you like about geography?
A – We see a lot of slides and things. I like the big pictures you get on the screen, they're really good – you think you're almost there.
Q – Have you ever listened to a tape recording?
A – Oh yes. We had one on villages the other day, it was really good. (This was a home-produced tape, acting out an interview with a Saxon Chief in order to explain the principles of early settlement location)
Q – Would you like to do more of these things in geography?
A – Yes.
Q – Does this sort of thing help you to understand more?
A – Yeah. I got bored just listening all the time.

Clearly a satisfied customer? But I contend that there is more to the use of educational technology than simply improving the pupil's ability to learn. I believe that the contemporary pupil is more sophisticated and demanding than his predecessor of some 25 years ago.

Increasing affluence over the past 30 years has seen a gradual dissemination of consumer high technology to a greater proportion of the population. The most recent innovation of the micro-chip, with advances in micro-circuitory and memory capability, is really only the latest manifestation of what has been a continuing trend. The child in our secondary schools is very much aware of these technical changes - indeed, many of the earliest spin-offs of the electronic revolution were in terms of sophisticated electronic games and cheap hand calculators - which they all want, and many use. The individual child is also being continually made more aware of the potential excitement of using that technology by considerable media coverage, almost on a daily basis.

It is not surprising, therefore, that the great majority of pupils now expect a great deal more from their teachers - in a way, it can be considered that the teacher is either competing with, or certainly being measured against, the technological excitement of the child's private social and cultural life. In simple terms - the pupil expects to see the use of modern technology in his own lessons at school. Much of the technology that may seem novel and strange to the teacher, pupils see as relevant and familiar.

This is a formidable challenge for the modern teacher to take up, and will certainly put to the test the degree of professionalism that he enjoys. As geography teachers, however, we are in a particularly strong position to be able to use much of modern technology to increase the variety of the approaches to our subject, and to amplify our powers of explanation.

But if we are to make easy and efficient use of technology in the classroom, it is essential that items of hardware (such as slide projectors, cassette recorders, television sets, etc.) are conveniently stored and readily accessible to every member of the department, and that items of software (such as slides, audio tapes and video cassettes, etc.) are not only accessible, but also adequately cross-referenced and maintained. A degree of thorough organisation is necessary if teachers are to be encouraged to use technology effectively. Much of this must rest upon the ability of the head of department to manage both resources, and colleagues effectively. Fortunately, more and more local authorities are now recognising the need for technical assistance in the geography department, which would relieve some of these pressures.

Nonetheless, the imaginative take-up of technology by the teaching profession as a whole can still be said to be rather slow. Consider, for example, the use of computing facilities in schools. For a surprisingly long time now, a large number of schools have possessed a computer, often quite powerful ones. However, the great limitations have been the complexity of operation, limited output facilities (usually only a print-out sheet) and the problem of access by departments other than mathematics, to computer time.

Today, however, micro desk-top computers with adequate computing power can be bought for between £600 and £1 300 (1980 figures), which should be within a number of departmental budgets. Such computers display information visually on a screen, and have a simple keyboard input operation.

In geography, the potential of such a machine goes far beyond the mere storage and retrieval of information. At all levels of the secondary school, it can be used to operate much more realistic simulations than can be handled adequately on paper alone, and to a degree of sophistication that allows the simulations to operate more closely to reality. This increases the interest and confidence of the pupils.

Further - even at a straightforward level, the computer can be used to demonstrate the validity of models, and illustrate them working, perhaps with the use of 3D diagrams. For sixth form students, the ability to handle more data in statistical exercises means greater confidence and reliability of the final result - and also an end to the unnecessary computation of bulky information on hand-calculators. There is also an increase in the number of services offered to geography departments using mini-computers, and ready-made programs can be purchased from GAPE for example (Geographical Association Package Exchange). Within my own local authority, Hertfordshire, the take-up by schools of mini-computers is already about one per week.

There is one other point which is rarely stated. It is most likely that the society into which current pupils will mature will be very computer-orientated, and despite the now feeble mutterings of a counter-culture, very technological. Already it is possible to buy toys for the under fives which require basic programming techniques for successful operation. How, then, can teachers ignore the micro-chip revolution?

The current demands being made upon us in the use of sophisticated technology are a challenge to our professionalism. We must critically examine the possibilities of each piece of hardware, and where relevant, put it into imaginative use inside the classroom. There is no doubt that such technological applications in teaching would be seen as 'obvious' by the pupils, in any case.

Historically, change has always come from the frustrated younger members of a society. In the teaching profession, and in geography teaching in particular, young teachers are enthusiastic about the potential of educational technology, but are all too often frustrated in its use. Will this frustration lead to a more professional attitude towards technology, or will interest be stifled before change can take place?

References

1 UNESCO (1965) *Source Book for Geography Teaching*. Longman (preface)

2 UNESCO (1965) *Op. cit.*, p. 75

Further reading

DES (1978) *Mixed Ability Work in Comprehensive Schools* (chs 3–4) HMSO

DES (1978) *The Teaching of Ideas in Geography*. HMSO

GAPE (1978) *Geographical Association Package Exchange*. Department of Geography, University of Loughborough, Leicester
GRAVES, N. J. (1978) 'Changes in attitudes towards the training of teachers in Geography', *Geography*, April, p. 75
GRUMMITT, S. J. (1980) 'The computer in the classroom', *Classroom Geographer*, January, p. 13
JONES, S. M. (1976) 'The challenge of change in geography teaching', *Geography*, November, p. 195
MILNER, S. (1978) 'Some pupil attitudes and the practical application of new geography in schools', *Classroom Geographer*, November, p. 15
SIXTH FORM WORKING GROUP (Geographical Association) (1979) 'Skills and techniques for Sixth Form Geography', in *Geography*, January, p. 37
STEVENS, S. W. (1980) 'A computer in the classroom?' *Classroom Geography*, February, p. 3
TAYLOR, W. (1980) 'Education in the eighties', *Geography*, January, p. 10–11

---

12 New opportunites in environmental education

Eleanor Rawling Associate Director, Schools Council 16–19 Geography Project

Environmental education and the geography teacher.

Are geography teachers doing enough to enhance the environmental education of our school students?

This might seem a provocative statement to direct at those teaching a subject which has always claimed as a prime concern the study of man-environment relationships.

The conclusions of a recent report seem to give some cause for doubt, however. The report, *Environmental Education in Urban Areas* (HMSO[1]), suggests that 'our education system has made people passive towards the built environment', lacking 'training in visual awareness and decision-making', and it states quite clearly that 'environmental education in secondary schools should be given a much higher priority than at present'.

The report is the result of the deliberations of a Working Party chaired by Professor Peter Hall and set up by the Environmental Board of the Department of the Environment. After an exhaustive survey of current environmental education activities across a range of ages, subject areas and institutions, the Working Party became convinced that there are large gaps in provision. Whilst the report was not concerned primarily with geography teaching nor only with the education of school children, its recommendations and conclusions should strike a note of concern to geography teachers everywhere.

Where are we going wrong? What else can be done through the medium of geography teaching? And why should it need the Department of Environmment and not the Department of Education and Science, to point out our weaknesses?

The answer to the last question – the involvement of the Department of the Environment – lies in the increasing concern felt by the environmental professions at their apparent failure to involve the public in environmental decision-making. As far back as 1969, the Skeffington Report[2] encouraged planners to formulate policies and plans on the basis of public participation, and indeed the need to obtain such public involvement has been written into subsequent planning acts and regulations. Despite these attempts, many planners feel that there has been little to show, in terms of improved decision-making or better planner–public relations. A large proportion of the public seems either unwilling or incapable of expressing views about the environment, or of making decisions with regard to its future. The British public appears to be lacking in 'environmental awareness'. For the planning profession and the Department of the Environment, this is a worrying feature, particularly at a time when many of the major planning priorities in the UK, such as reversing inner city decline and managing the urban fringe, require positive co-operation and input from the general public. Thus, the Department of the Environment was really acting in its own interests when it set up the Working Party, and was making a constructive approach to the teaching profession when it ultimately recommended more action in school education.

The type of action needed was made quite explicit in the report. It pointed out that environmental education is not just a matter of environmental knowledge nor is it only obtained through academic study. It is concerned equally with skills, and with feelings and emotions, and it *must* involve direct experience. As far as geography teaching in schools is concerned, it seems that we should re-examine our teaching programmes and teaching approaches. Is it perhaps the case that we are over-concerned with inculcating knowledge *about* the environment and too little involved in providing children with direct experience *in* the environment and with using the environment as a *medium* for learning skills? Are we too busy teaching models and theories of urban structure to be concerned whether our students have a viewpoint about their own town or are able to take part in shaping its future?

Any school geography course requires a balance in terms of content covered and skills developed. However, it may be that most secondary school geography courses could give a greater emphasis to local environmental work. The Hall Report stated unequivocally that in its view 'the emphasis should be on local studies, with locally produced materials'. It would certainly seem that there are many local issues of a type which would lend themselves to geographical enquiry – the location of a new shopping complex, the redevelopment of the town centre, village growth in the urban fringe – but where does one start in finding out about such issues? Which are the best issues to choose? What resources exist? Who are the key decision-makers?

One answer was provided in the Hall Report, which commented that 'one of the most important new moves in developing the skills of decision making and participation has been the involvement of some planning departments in education'. It went on to recommend that environmental work should build on the potential for bringing educational and environmental professions together. Planning is not the only environmental profession with interests in the education system or with contributions to make a local issue-based work. The potential exists for geography teachers to draw on the knowledge and skills of a variety of environmental professionals in their own local area – architects, surveyors, estate agents and local community workers, to mention but a few. It is likely, however, that, partly as a result of initiatives from planners themselves, and partly because of the planners' special concern with land use and the management of space, it is in this area that geography teachers will discover the greatest possibilities for co-operation.

The potential for joint work is undoubtedly great. Individual initiatives in teacher–planner co-operation in a number of different areas of England and Wales are beginning to show the variety of activities possible. They range from sharing of resources and data to direct involvement of planners in running student projects, and even to school students joining in with the work of a planning department for a day.

Before the value of such work is discussed, it may be appropriate to look at an example – one of the results of teacher–planner co-operation.

Wantage 1991

'But who decided on the growth options for this part of Oxfordshire in the first place? Shouldn't the public be drawn in at an earlier stage in Local Plan preparation'.

'Well – er – you may be right in one sense, but there are certain unavoidable constraints which limit building in the Wantage area, and it certainly saves time in the plan-making process if the planning department can point these out first.'

So ran the dialogue, as the planner defended his department's procedures, not at the Council chambers in south-west Oxfordshire, or at an awkward public meeting in Wantage, but in a sixth-form classroom at a comprehensive school in Abingdon.

Participants in the debate were members of the lower sixth geography set, who were completing their involvement in a Local Plan simulation by questioning the local planner who had jointly prepared and run the exercise with them during the previous two weeks.

'Wantage 1991' involves students in the preparation of a local plan for the small market town of Wantage in south-west Oxfordshire. The complete sequence of activities covers the main processes of plan-making and so begins with problem-definition and data collection, proceeding through analysis, determination of constraints and objectives and testing of alternative plans, to plan evaluation and decision making. It will be realised that this also provides a well organised route for enquiry learning.

The problem posed for the students, through use of local newspaper cuttings and an extract from the Structure Plan,[3] is that of the future growth of Wantage. Together with its neighbouring villages of Grove and East Challow, the present town of Wantage comprises a population of 15 000. However, Structure Plan statements propose that the town should grow by 7 500 by 1991. This large population increase will also necessitate an increase in employment of 2 500 jobs, a new shopping precinct and improvements to the east–west road system. The questions are: 'How will Wantage cope with such growth? Where will the new areas of housing, employment and shopping be accommodated? What impact will such growth have on the lives and environments of existing inhabitants?'

The first stage of the exercise comprises analysis of Ordnance Survey plans and maps at various scales and of aerial photographs borrowed from the Local Planning Department. The students themselves attempt to identify both constraints and opportunities for the Wantage area. In so doing, they prepare a list of those locations which they should study at first hand on a site visit. Preliminary talks with both the teacher and the local planner result in the establishment of a suitable itinerary, and the site visit then takes place by minibus. On the several occasions that this exercise has run, both planner and teacher have accompanied the students, sometimes together and sometimes separately. Plenty of time is allowed for students to walk around, sketch, take photographs and talk to the local Wantage people.

On return to the classroom, students record their observations and attempt to set out clearly the potential areas for growth in Wantage.

The next stage in the exercise is the role-play of the Local Plan meeting. The scene is set in a series of imaginary letters written to representatives of the District and County Councils and various influential pressure groups in the area (see Fig. 34). It is assumed that the South-West Oxfordshire Planning Department have carried out a technical appraisal of Wantage and the surrounding area and identified some forty-four parcels of land capable of development, of which some fifteen are required for residential development, a further three for employment purposes and one as a new shopping centre (see Fig. 35). Three alternative road lines have also been

Wantage Chamber of Trade

Chairman: William Bunce,
45 Charlton Avenue,
Wantage,
Oxfordshire.

Dear Joe,

Just a note to remind you that you agreed to represent the Traders of Wantage at the meeting with the Planners. As we've both been in business in the Market Place since the War, I need hardly remind you how important it is that Wantage keeps growing. We must tell the Planners that its no use sticking all the houses out at Grove, the new-comers are as likely to take their trade off to Abingdon or even Oxford, as they are to drive into Wantage, particularly with all the parking restrictions they keep forcing on us. I've even heard a rumour that the Co-op, or someone, is interested in building a hypermarket at Grove if there is much more growth there. No, we must have more growth, and it must be right on the edge of Wantage where people can easily get to the Town Centre. I think the area just south of town would be ideal, can't understand why it hasn't been developed years ago. Suppose those interfering conservationists are standing in the way of progress again - most of them don't even live in Wantage!

The other thing we ought to press for, is the Inner Relief Road. The Market Place won't stand any more through traffic - its even started to kill trade - but on the other hand we don't want to divert it all onto a By-Pass which would miss the town altogether. I understand that the full By-Pass would be more expensive anyway, and you know who will have to foot the bill as usual - the poor old traders who seem to pay most of the rates in this town.

Yours,

Bill.

William Bunce.

To: Mr Joseph Soap,
Treasurer of Wantage Chamber of Trade,
Soap's Discount Store,
2-4 Market Place,
Wantage.

Fig. 34

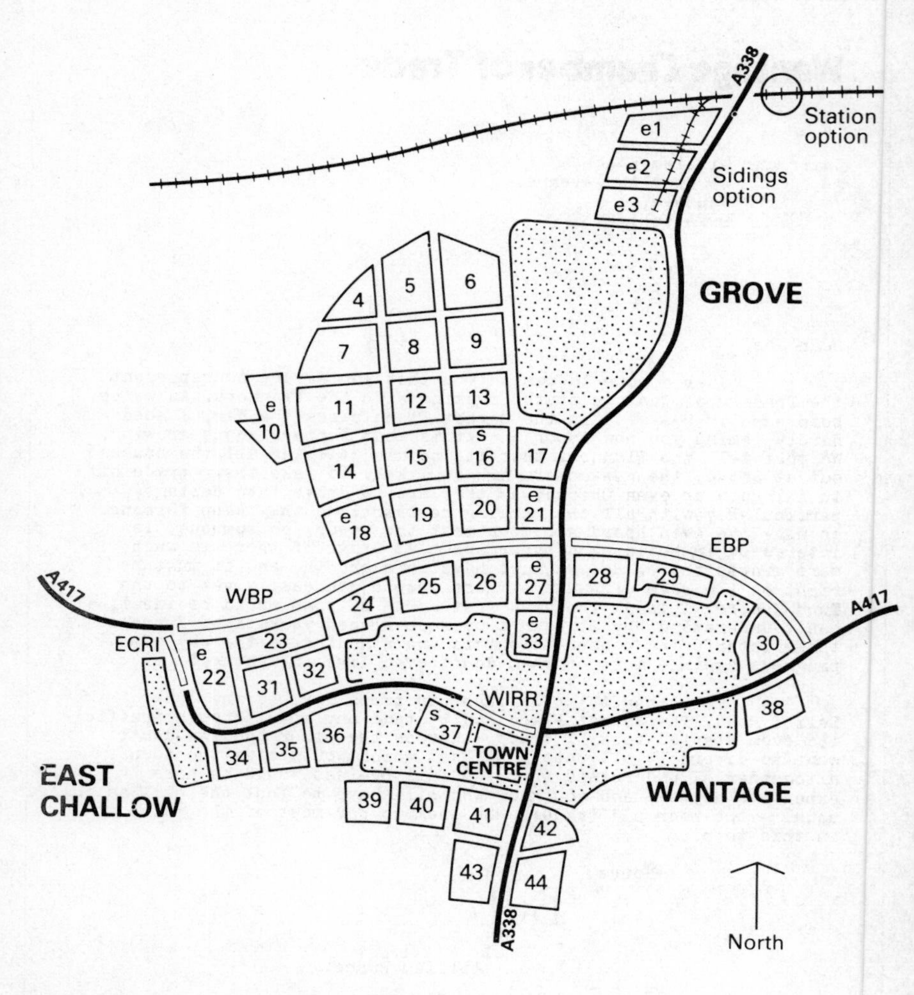

Residential: Select 15 out of 44 units
Employment: Select 3 out of 9 units (e)
Shopping: Select 1 out of 2 units (s)

East-west road: Full by-pass (WBP + EBP) = £1,400,000
Partial by-pass (WBP only) = £ 850,000
Inner relief system (WIRR + ECRI) = £1,150,000

Fig. 35 *Plan of the forty-four parcels of land capable for development.* (*Dotted areas = existing built-up areas of town.*)

identified, each of which results in different financial and environmental costs. Options regarding re-opening the station and providing rail sidings for industrialists have also been identified.

So far, the District Council have already run one lively but inconclusive public meeting to discuss the Local Plan. As a result, the Planning Committee have authorised the Chairman to meet representatives of the various interest groups as being the potential major objectors to the Local Plan proposals. These include the British Road Federation, the Council for the Protection of Rural England, the Friends of East Challow, Grove Parish Council, the National Farmers' Union, the Wantage Chamber of Trade and Wessex Homes Limited (a major local house-builder). The Chairman of the County Highway Committee has also been invited as it is understood that there is some reluctance on the part of the County Council to finance some of the more ambitious highway proposals in the Draft Local Plan. The objective of the meeting is to reach agreement about the broad outlines of a future plan for Wantage.

Each student receives a briefing letter (similar to Fig. 34), setting out in perhaps rather extreme terms the major objectives of the particular pressure group which the representative should safeguard at the meeting. The characters are designed to be slightly 'larger than life' in order to highlight the conflicts which exist within the community.

The Local Plan meeting is enacted in the classroom and usually requires a session of one and half to two hours. One meeting comprises seven to nine participants, so that in order to run at least two meetings for comparison purposes it may be necessary and valuable for the teacher and planner to join in. If this is done, experience shows that it is better for them to play a normal role rather than attempting to direct proceedings. In most cases the debate is lively. Indeed, the Chairman requires a strong personality to bring about any form of consensus plan. At the end of two hours, it becomes increasingly obvious to the students that the various interest groups hold a range of conflicting views about the future of Wantage. There is not one ideal solution; political bargaining and negotiation are the only way to reach a satisfactory compromise.

With the meeting over, the task for the students is now to carry out some sort of evaluation of plans produced by the different role-play groups. Evaluation can become a complete separate exercise if time is available, using perhaps some form of Planning Balance Sheet and assessing each plan against a series of objectives which the groups have identified (e.g. need to protect the Area of Outstanding Natural Beauty; desirability of removing traffic from Conservation Areas; minimisation of journey to work distances; protection of prime agricultural land, etc.). The objectives usually fall into three main categories: namely, those concerned with economy, those to do with convenience and those emphasising environmental quality.

Students quickly find that whilst it is possible to decide which is the most economic solution, or the one most convenient for the town's inhabitants, or the one which does least damage to the environment, it is rarely possible to discover a single plan which best satisfies all objectives! Planning is revealed again to be a process of compromise between alternative priorities.

The final stage in the exercise is that of decision making. Having recognised the difficulty of the task set, students are now placed in the role of decision maker. 'As Local Planner for the Wantage area you are aware of the range of conflicting feelings about the town's future. Nevertheless, a decision must be made and a future plan for Wantage presented to the Planning Committee. What is it to be?'

In reaching this final decision, students must necessarily clarify their own attitudes about this local town, and attempt to find a well-justified compromise solution. Of course, it is at this stage that many students begin to question the whole process of local plan making. Hence, the opening dialogue; for it is usually found valuable to invite the planner to be present at the session in which decisions about Wantage are discussed. It provides a stimulating, if sometimes uncomfortable, hour for the planner!

The value of the exercise

It must be remembered that 'Wantage 1991' presents an issue which was topical and local to the students concerned. By the time this chapter is printed, another issue will provide a more relevant basis for local geographical enquiry in south-west Oxfordshire. In other parts of the country, teachers must choose their own local issues if the work carried out is to achieve its objectives of introducing students to local decision making and of developing genuine environmental awareness.

There is still value in analysing the Wantage exercise, however, since it provides one model of issue-based work, and a general reminder of the benefits of teacher–planner co-operation.

'Wantage 1991' presents students with a thorough introduction to the Local Plan process as it operates in their local area. This is useful knowledge, but it could have been imparted through lecture or notes, if this was to be the main objective of the exercise. More important is the opportunity provided for students to get out into the local area to examine an issue at first hand, to become aware of the place and the people and to meet the previously 'face-less' planner who deals with the town. In this way students begin to achieve valuable understanding of the local environment and the local decision-making system, within which many of them will play a part.

Since it was constructed as a geographical enquiry in the first instance,

'Wantage 1991' also reveals the way in which skills and techniques used by geographers can be included in local environmental work. Students are involved in map and photograph analysis, sketch plan presentation, analysis of simple statistics and streetwork observation.

The role-play element of Wantage 1991 helps to reveal the conflicting perceptions and viewpoints which individuals hold about their town. Students appreciate the difficulties implicit in environmental decision making and the necessity for compromise. They are forced to clarify their own attitudes towards the local area, and to realise the implications for others of the views they hold. The heated discussions about the draft Wantage Plan show that students quickly learn that environmental facts are dry and meaningless without environmental feeling and environmental commitment.

It would be hoped that such enlightenment carries over into life beyond school. Judging by the fact that some students involved in Wantage 1991 have also followed up their interest by attending a local meeting or joining in a participation exercise, such 'spin-offs' do seem possible.

The exercise 'Wantage 1991' was constructed jointly by a planner from the Vale of White Horse District Council and a geography teacher from John Mason Comprehensive School, Abingdon. The co-operation began as a result of mutual interest in the local area, and a recognition of the advantages to be gained. The borrowing of resources from the Planning Department and the occasional visit of a planner to a school geography lesson eventually resulted in the establishment of co-operative work on a more formal basis, with the agreement of both the Headmaster and the Chief Planning Officer. At the present time, the Vale of White Horse possesses a flourishing Environmental Education Group, comprising a mixture of teachers and planners. The 'Wantage' exercise is now merely of historical interest to this group, whose energies are channelled into the production of regular 'issue sheets' and of teaching exercises based on local topics. Their work does, however, build on the foundations laid by 'Wantage 1991'. A good piece of local environmental enquiry makes use of the planner's knowledge of the planning process and insight into local area politics. It will draw on the maps, plans and statistics available at the planning department and build into its sequence any relevant local meetings or public inquiries. The teachers represented in the group will ensure that the geographical implications of the issue are drawn out, that geographical skills and techniques can be practiced where possible, and that appropriate teaching approaches are employed. Thus, teacher–planner co-operation draws on the skills and expertise of each professional to the benefit of students.

The Wantage exercise and the Vale of White Horse Environmental Education Group are but one example in an area of increasing opportunity. Stimulating environmental education work is also being carried out, for instance in Avon, where strong links have developed between the County Education and County Planning Departments; Hammersmith, where valuable work is being established through the efforts of a School Liaison Officer in the Planning Department; and Nottingham, where a project was set up involving co-operation between the Planning Department, Trent Polytechnic and local teachers.[4] In many of the most successful cases, local environmental work begins through small-scale initiatives by interested individual teachers and planners. It may be that such ventures are most likely to succeed in the present time of economic restraint. Formal recognition and more adventurous activities may then be forthcoming once the work has been established.

The planning profession has undoubtedly realised that the time is ripe for co-operation with the education world. The Hall report quoted at the beginning of this chapter reveals the importance with which many planners now view environmental education. In addition, the Royal Town Planning Institute, following on from the lead provided some years ago by the Town and Country Planning Association, has now set up an Environmental Education Working Group, committed to supporting and encouraging co-operative planning and education ventures. A booklet recently published by this group, *The Role of the Planner in Environmental Education*[5] makes clear to planners the way in which their contribution can be most valuable to teachers.

It seems, then, that it is time for teachers of all environmental subjects, and not least geography, to grasp these new opportunities for co-operation with planners and so to ensure that the subject does fulfil its potential in environmental education.

References

1 HMSO (1979) '*Environmental Education in Urban Areas*'. Report by a Working Party of the Environmental Board of the Department of the Environment, chaired by Professor Peter Hall.
2 HMSO (1969) *Public Participation in Planning*. Report of a committee set up by the Department of the Environment and chaired by Arthur Skeffington.
3 The Structure Plan statements used in this exercise were based on actual proposals, but were not taken directly from the Draft Oxfordshire Structure Plan.
4 For other examples of work involving teacher–planner co-operation, see *Bulletin of Environmental Education* (93), January 1979; and *The Planner* **65** (4), July 1979.
5 ROYAL TOWN PLANNING INSTITUTE (1979) *The Role of the Planner in Environmental Education*. A booklet produced by the RTPI Working Party on Environmental Education.

13 Evaluation in the 1980s

Gerry Hones University of Bath

It seems probable that the next decade will be a difficult one for teachers as they cope with a number of changes in their working environment. Some, such as the effects of economic cuts or falling school rolls, are outside their control but others are more likely to involve the teacher in decisions concerning the rate and scale of the change. For the geography teacher this will naturally include the effect of changes in the subject itself and related teaching methods.

To suggest, therefore, that geography teachers should at the same time also make a careful review of their position as 'evaluators' may seem ill-timed, unrealistic and highly impractical! Yet there appears to be a case for changing their evaluative role in the near future, in spite of all the problems that this could create. How far individuals will accept the argument, at least in principle, seems likely to be strongly influenced by their views on:

(a) the place of geography in the total curriculum;
(b) the procedure that should be used for *certification* of their pupils; and
(c) the growing demands for some form of *accountability* process.

Particularly important is the degree to which teachers see themselves as being directly involved in the latter two processes, with all the extra commitment and responsibilities that would naturally ensue.

Of course, many teachers are already deeply involved in the significant moves towards increased teacher participation in the certification process in recent years – for example, assessing units of work as part of a larger framework or designing and operating complete Mode 3 examinations. As a result they know something of the delights and problems involved! Nevertheless, seen in world terms, the changes have been relatively slow and much less complete (Broadfoot,[1] Maguire,[2] Neave,[3]). Why is this so? Do peculiar circumstances apply to this country?

Clearly the procedure whereby the certification of pupils is wholly, or even mostly, in the hands of agencies outside the schools is seen as strange by teachers in many countries – for example, Canada and Australia. It is worthy of note, however, that in one Canadian province studied recently, the teachers pressing for the freedom 'to teach and to *assess* their own students according to their professional judgement (Lawton,[4]) were at the same time accepting that there should be 'wider agreement about the whole curriculum'. Relating this to what he sees as a clear pattern of moves in this country towards teachers yielding control over curriculum planning, Lawton argued that teachers should increase their influence in the evaluative procedures, building this into a system of accountability.

Not everyone would agree that 'evaluation can make accountability an equitable, supportive, nonpunitive process' (Hayman and Napier[2]). Nevertheless, it certainly seems possible that, by taking a more central role in the certification of their pupils, geography teachers could turn the needs of accountability to their own advantage. If the accountability is to the public – thus including both parents and prospective employers – then this could be achieved by demonstrating the value of the geographic education received by the pupils through the medium of an imaginative evaluation process.

Instead of a pupil's certification being represented by a single grade, the opportunity would be there for the whole process to be opened and widened. The professional teacher would be accepting the responsibility of explaining the system to the non-professional public – including the means by which the final gradings were arrived at. An assessment 'profile' for each pupil's work in geography could provide a more useful framework of information about that individual. This appraisal would not need to be restricted to the cognitive aspects and could surely include wider social skills and some attempt to consider the affective domain. The ability of a pupil to work as member of a group, to analyse a problem, and to present a case orally, are all possible components of such a profile.

In justifying a number of subjective judgements which formed part of the evaluative process, it is also likely that the teacher would be explaining what comprised the geography curriculum, why it was included and how it was developed, both in and out of the classroom.

An explanation of why pupils had been studying the way in which people reacted to the ever-present threat of such natural hazards as earthquakes or volcanoes – or the way individual pupils set up their own fieldwork exercises to measure stream flow variations or differing perceptions of a neighbourhood – or the value of groups of pupils being asked to produce their analysis of a local planning problem – could surely offer the opportunity for a much better understanding of the value of the work in process.

All this would call for considerable evaluative expertise as well as confidence in the value of what was being taught and evaluated. Until there was adequate support from the authorities, in terms of the necessary resources and opportunities for in-service training, it would probably be best if moves towards greater teacher control of evaluation were operated in stages. The first could be for teachers to accept a greater responsibility for the assessment of their pupils but within the overall framework of the examination boards – with the latter assisting in the formulation of evaluative items, and moderating the results.

But the key question remains for the 1980s: do geography teachers really want this degree of involvement in the assessment of their pupils, with all the pressures as well as opportunities that would develop?

References

1 BROADFOOT, P. (1979) *Assessment, Schools and Society*. Methuen
2 MAGUIRE, J. (1976) *An Outline of Assessment Methods in Secondary Education in Selected Countries*. Scottish Council for Research in Education
3 NEAVE, G. (1980) 'Developments in Europe', in Burgess and Adams chapter 7, pp. 72–89
4 LAWTON, D. (1980) 'Responsible partners', *The Times Educational Supplement*, 7.3.80, p. 4
5 HAYMAN, J. L. and NAPIER, R. N. (1975) *Evaluation in the Schools: A Human Process for Renewal*. Wadsworth, Preface, p. viii

Further reading

BURGESS, T. ADAMS, E. and (eds) (1980) *Outcomes of Education*. Macmillan
STRAUGHAN, R. and WRIGLEY, J. (eds) (1980) *Values and Evaluation in Education*. Harper and Row

---

14 Language, ideologies and geography teaching

Rex Walford University of Cambridge

> 'When *I* use a word ... it means just what I choose it to mean'
>
> HUMPTY-DUMPTY *Alice Through the Looking Glass*

We have lived through a decade in which many documents about education have been published and there was once a time when I was innocent enough to believe that they all actually meant what they said. It was a touching, and, as I see now, fatally misplaced, belief that the act of authorship implied commitment to clarity. I do not mean to suggest that the authors of academic books, position papers and official reports are counterfeiters; only that there is a tangled web which ultimately deceives.

In educational discourse, there are traditional obeisances made in disarming fashion. I have not yet come across anyone publicly opposed, for instance to such objectives as 'the development of the full autonomy of the individual' and 'a commitment to rigorous investigation' though I have been in classrooms where one or other of these objectives has been pursued to the total exclusion of the other.

In the same way, a stated concern for 'aims and objectives', for 'flexible and varied classroom strategies' and for 'sensitive evaluation' seems a compulsory doxology for any article which describes a classroom experience. Whether these aims are put in to practice has often to be taken on trust.

However, this language is an entrée to a world in which intellectual gatekeepers require verbal visas to be shown by those who seek to enter.

I labour this initial point, because it seems to me that it is vital to recognise that there is sometimes a secret language of many documents which is not at all the same as its surface one. Those who are most skilful at this game manage to offer a multiplicity of interpretations and meanings within the same statement, so that any number of readers can be satisfied by it. Looking for particular safe 'cue-words', an individual reader is reassured to find the evidence for which he is looking somewhere in the statement, and thus goes on to imagine that his own view is the dominant force in the thinking of the writer – when it may be nothing of the kind.

Such writing appears in the bland profession of aims in school prospectuses, in some 'official' publications and in the halting justificatory statements required of subject departments in school curriculum discussion. An 'all things to all men' approach forestalls criticism and straddles internal disagreements.

Some of this material is knowingly anaesthetic since the formation and discussion of educational policy in linguistic terms is often unrelated to the day-to-day realities of the classroom. The soothing balm of well-loved phrases used in familiar conjunction with each other can be both therapy and inertia.

I do not know if those who draft official reports use the *lingua franca* of educational jargon knowingly in order to sweeten bitter pills, or unknowingly in the optimistic belief that it will be understood. But even if readers recognise the language and subsequently use it, that is no guarantee of its being understood, nor of its effect on years of ingrained practice.

Take, as an example, the definition of education used in the Warnock Report and quoted with approval in the HMI pamphlet, 'A view of the curriculum' (1980)

> '... first to enlarge a child's knowledge, experience and imaginative understanding, and thus his awareness of moral values and capacity for enjoyment; and secondly to enable him to enter the world after formal education is over as an active participant in society and a responsible contributor to it, capable of achieving as much independence as possible ...'

All human life is there. But examine the statement more closely.

In the first proposition; is the 'thus' (line 2) a clear and *undoubted* conjunction between the first and second parts of the sentence? Is it true that 'enlarging a child's knowledge' *necessarily* increases his or her capacity for enjoyment?

Will enlarging the child's experience and imaginative understanding *necessarily* bring an awareness of moral values? If so, is the choice about acting on the values to be left to the child, or is there a hidden agenda which the statement does not specify?

In the second proposition, if the child is to enter the world as an 'active participant in society', is it acceptable to make him or her sharply critical of a ruling government's policies in being 'responsible contributors to it'? Or is political judgement *off* the agenda? If children are to achieve 'as much independence as possible' through school what if that conflicts with views which their parents may hold about the need to maintain strong filial ties? (a particularly keen issue with many immigrant families).

There are, no doubt, elucidations and answers to be found, if only the original authors would spare some time to be subjected to a cross-examination of their statements – but that is hardly the point. The statement itself, in seeking to command a wide agreement, makes large assumptions and contains inherent contradictions. Put it another way; it doesn't really say anything at all....

The same is true of some statements made about geographical education. What is one to make, for example, of the following set of aims which are described by a school geography department:

> 'In studying geography pupils come to have a general knowledge of the world in which they live. Topics in both human and physical geography will be considered and important ideas and themes will be covered by a variety of approaches, including fieldwork. Pupils will come to have a wider appreciation of important environmental issues and the subject will be useful to them in equipping them with skills for a wide range of careers....'

Such a statement says everything, and it says nothing – it gives little clue about what is really going to happen in geography lessons in that school although it pays lip-service to most of the in-phrases which are regular currency within the discipline.

What lies behind statements of these kinds is another level of reality; a level on which particular ideological positions are usually held (either implicitly or explicitly) but which are *not* revealed in the public language of justification. Such positions may be based on experience, or 'what we feel is right in our bones' or genuine and democratic discussion with colleagues or on the exigencies of an examination requirement; but they rarely, in practice, seem to have much to do with written statements. Indeed there are a number of case-studies which reveal that general statements are placed on top of course descriptions after the main work or practicalities has been decided; the cherry in the cocktail, so to speak, making it altogether more alluring, but without any relation to the liquid ingredients.

The ideological positions are visible only accidentally and intermittently but they form the strong basis of many teaching strategies. What follows in this paper is an attempt to speculate about some of the general ideological traditions which affect British education and to define styles of geography teaching which relate (knowingly or otherwise) to them.

The 'liberal humanitarian' tradition

Those who view education as primarily concerned with the passing on of the cultural heritage from one generation to another reside in this tradition. The distinguished line of advocates stretches back to Plato, and includes Matthew Arnold and T. S. Eliot; among modern educational writers G. H. Bantock represents an eloquent expositor of the view.

In this tradition the importance of maintaining a continuity of worthwhile ideas is seen as being of paramount significance and thus there is a defined curriculum which students *ought* to be taught. Such a 'map of knowledge' is usually marked out in strong subject disciplines though it may also be justified by theorists in terms of distinctive ways of learning, or areas of experience.

Teachers are seen as the guardians and gate-keepers of the jewels of civilisation and their role is to initiate children into an appreciation of them. Whether *all* children are capable of understanding and/or benefiting from such an education is a matter of some discussion within the tradition; both Eliot and Bantock, for example, speak of one culture which represents 'high' or 'elite' learning and another 'folk culture'.

It is tempting to hypothesise an echo of the 'two cultures' view of this tradition in the original institution of *two* Schools Council funded projects in geography in 1970. Was Bristol to develop geography in a 'high culture' tradition and Avery Hill to work in the other sphere? The subsequent history of the Projects belies the hypothesis since the Projects worked closely together and were different in their strategies, rather than in their views of geography.

However, it could be argued that much of the effort to give geography a strongly conceptual and rigorously scientific outlook through the tools of spatial analysis was influenced by the desire to give it a worthy place in the 'cultural heritage'; a place in the sun, so to speak, alongside its older and somewhat more highly-considered companion, history. The development of geographic curricula in selective schools is certainly of this kind, and the spatial tradition in schools has led to a systematised, conceptual framework in some syllabuses which represents a conspicuous 'intellectualisation' of the discipline.

The child-centred tradition

It is also possible to speak of the educative process in terms of the self-development or the bringing-to-maturity of the individual student. In this tradition, the need for the pupil to discover self-autonomy and social harmony is strongly valued, and such a 'child-centred' tradition has strong support from many psychologists of childhood and from educators who have been involved with the development of very young children.

The best of British primary school tradition (as exemplified by Susan Isaacs and her followers) follows this ideology, which derives also from the work of Froebel and Pestalozzi. Rousseau sought a similar objective in his individual education of his pupil, Emile; John Dewey, on the other hand, emphasised the social skills and democratic experiences which a child needed to experience with others in order to fully mature.

Within this tradition there is a valuing (almost mystical at times) of direct contact with the outside world and a cherishing of subjective experience. Childhood and adolescence are regarded as states worthy in themselves and to be enjoyed for their own sake. There is more emphasis on the *process* of education day-to-day than on the product in terms of long-term goals.

Geographers who have been influenced by this tradition do not, in all probability, care very much whether they are 'geographers' or not. They would see their role primarily as an educator of the whole person, breaking down the artificial barriers of 'subjects' and integrating experience wherever possible. Geography itself is only important in so far as it helps a child towards self-understanding.

Thus, an important geographical contribution to this kind of tradition would be exploration of personal feelings and images about the world, and the orientation of the student in his own personal, neighbourhood and regional environment. The experimental field work of the kind pioneered by Brian Goodey, Jeff Bishop and the Town and Country Planning Association Education Unit (see BEE June 1980) would be valued highly.

Individual work and projects would be encouraged, with a strong predilection for the student rather than the tutor to have the final word in defining the topic. In the classroom, there would be concern for work to be self-motivating, even if the teacher intervenes by providing an environment in which useful 'enquiry learning' may conveniently take place.

The utilitarian tradition

Those who see the main job of education as preparing pupils to go well-equipped into society are the inheritors of a utilitarian ideology in education. The assumption is made that the role of the school is to help the pupil survive in an already-defined situation, and that the curriculum therefore needs to be primarily geared towards the provision of skills and knowledge which are useful in helping him or her to 'get a job' and 'earn a living'.

State policy on education has often been dominated by such an outlook, and it is possible to cite examples in places as diverse as Japan, Stalinist Russia and Hitler's Germany. Education is regarded as a tool of state or social policy; it can even be a dangerous activity if carried out too thoroughly or enthusiastically.

Since James Callaghan instituted the so-called 'Great Debate' about education in the UK in 1976, the implicit ideology in many official statements about education has been in the utilitarian tradition. Schools are seen to be failing society if they do not produce a work-force with requisite skills and demeanours to make industry successful; poor exporting performance is traced back to the deficiencies of second language teaching and a greater emphasis on resources for French teaching is seen as a contribution to redeeming the malaise. . . .

Subjects with a strong vocational significance are highly valued in this view of education and geography may come fairly low in the priorities. It may be argued that prospective lorry drivers will need to be equipped with map-skills and that travel-agent assistants will need to find their way amongst the package brochures – but the intellectual challenge of locational analysis of one hand, and the sensitivity of experiential fieldwork on the other is not likely to be held in great value.

It is more likely that the role of geography, seen within this tradition, will be defined in an adjunctive role; it will be informational and descriptive to some extent, designed to equip students to read TV bulletins without being bemused, and to maintain casual social conversations, but not to *be* geographers.

To some teaching the subject, it will be seen ultimately justified as another avenue to a paper qualification which helps to give the student the necessary entry to certain occupations or further courses of study.

The 'reconstructionist' tradition

A fourth possibility is that education is seen as a potential agent for *changing* society. In this ideological position, the task of the educator is to engender a kind of divine discontent in the student, so that the latter does not lightly accept things as they are.

The school will encourage challenge to the accepted status quo (competitiveness in capitalist societies, collectivism in the Communist world) and try to give students insights into alternatives, or into the possibilities for improvement.

This means that there may be clashes with parental, societal or media views, since the ideology by definition seeks to initiate a critique of the society in which the student lives.

Reconstructionist thinkers have ranged from social utopians (such as H. G. Wells and William Temple) to radicals (such as Paoulo Freire and Ivan Illich). Freire's basic literacy programme politicised rural villages so effectively that he was removed from his post and traditional ineffective approaches reinstated. Illich's reconstructionism extends towards an unbelief in schools themselves, and a preference for informal learning networks.

A geography based on 'reconstructionist' ideas might be expected to emphasise spatial injustices and imbalances in society and to have a strong desire to develop social and environmental concern in pupils. The open discussion of values and feelings would be given prominence; teachers would take up one of the attitudes to values education outlined by John Huckle in his chapter in Part Three. Active participation in decision-making exercises might be encouraged, and subsequent social action by pupils not ruled out (e.g. in relation to concerns in their own immediate environment, or in relation to issues about which they felt informed but frustrated).

Is it significant that the most visible signs of this kind of geographical education are in schools where the pupils are, for the most part, well-motivated, and relatively affluent, and where potential explosiveness is often gently defused by circumstance? If the same material was taught to some disadvantaged classes, it might have considerable repercussions and land the teacher in delicate situations. . . .

The polarised and abbreviated sketch of these traditions does, of course, make some simplifications in the cause of seeking distinctiveness. In practice, one hardly ever finds a teacher or department which works exclusively from one ideological base; nevertheless, it is also unlikely that one will find a totally 'balanced' geography programme – however much claims to the contrary are made.

In *practice*, I submit, many geographers in schools describe that work using the respectable innocuousness of the language which I exemplified earlier in this paper. But beneath, they shape their lesson-plans, their work-units and their syllabuses with one of these ideological traditions as a dominant factor in their professional work. It may be either explicit, or implicit.

The casual remark over staff-room coffee sometimes lifts the curtain for a moment. . . . 'Of course, the fundamental thing is to give them a grasp of the basic ideas of the discipline'. . . . 'I don't really care what they got from it, but the trip was a tremendous experience'. . . . 'We are really here to help them get an O-level aren't we – that's all'. . . . 'If they just understand how the system gets at them I'll be happy'. . . .

It is not the purpose of this paper to suggest which, if any, of these ideologies is the 'right one' – or even to ask if that question itself it proper. Readers will no doubt make their own deductions about where the author's own sympathies lie, since no writing can be totally free from personal bias and perception. (Elsewhere in this Part John Huckle makes a powerful plea for the reconstructionist ethic to predominate.)

One important contemporary issue is germane to the general discussion of language and ideologies which has gone on so far.

Recent government pronouncements about the curriculum reveal an implicit utilitarian approval towards education. Should, as Richard Daugherty suggests in Part Two, practitioners of geography manoeuvre gently into the prevailing breezes and sail before the winds of change which seem likely to influence strongly curriculum priorities in the eighties?

If so, whatever personal feelings are held about the nature of geography, the survival of the subject on school timetables may be best achieved by a return to a moderate informational tradition, laced with the general discussion of world energy resources, population trends, the risk of natural hazards in certain places and other 'relevant' issue-based material.

The alternative is to maintain a belief in a personal view and assume that the professional autonomy of the individual teacher will be strong enough to withstand whatever pressures emanate from the Head's study, the Adviser's office, the Inspector's visit, the Governor's meeting or the Green/White/Black papers of officialdom. Preoccupation with far horizons may cause the manoeuvrings of other subjects which respond to more immediate happenings to be overlooked.

Thus, a salutary thought for the next decade may be that geography's place as a school subject may depend more on survival strategies within curriculum planning than on its well-being as an intellectual discipline. But perhaps it should also be recognised that those educators conscious of their own intellectual roots will be much better placed to contribute to the ultimate shaping of education than those to whom this consciousness seems illusory or superfluous.